COLORING SOUTHERN CALIFORNIA
BUTTERFLIES & CATERPILLARS

BILL HOWELL

Sunbelt Publications, Inc.
San Diego, California

Coloring Southern California Butterflies & Caterpillars
Part of the *Color & Learn* Series

Sunbelt Publications, Inc.
Copyright© 2020 by Bill Howell
All rights reserved. First edition 2020

Cover and book design by Kristina Filley
Parts of a butterfly and caterpillar drawings by Rebecca Kriz
Project management by Deborah Young

Printed in United States of America

No part of this book may be reproduced in any form without permission from the publisher.

Please direct comments and inquiries to:

Sunbelt Publications, Inc.
P.O. Box 191126
San Diego, CA 92159-1126
(619) 258-4911, fax: (619) 258-4916
www.sunbeltpublications.com

23 22 21 20   4 3 2 1

All photographs are by the author unless noted.

Introduction to *Coloring Southern California Butterflies & Caterpillars*

Everybody enjoys observing birds, but butterflies have also intrigued us with their fascinating flight, captivating colors, and intricate patterns. With contemporary digital cameras and the latest close-up binoculars, butterfly surveillance is now rivaling birdwatching. Butterflies have a unique life cycle and wondrous stories to tell.

A coloring book is a stress-free way to initiate your learning about these frequent fliers. This edition allows a child to joyfully color butterflies and moths and not worry about staying inside the lines and encourages an adult to reconnect with the past. Better yet, how fun for a youngster and an oldster to color together. Each highlighted creature in this book has a write-up that includes information about the dominant hues, tints, spots, and dots on the wings. Also revealed are habitats and seasons when these big winged insects may be encountered and amazing facts about each species. You will become a better artist and add to your knowledge about each flutterby.

Some of these southern California butterflies, like the Painted Lady, have a worldwide range, while others, like the Hermes Copper, fly basically only in one county. Certain species are very unique, like the California Dogface. This State insect has a likeness of a poodle dog on each upper forewing. Another unique species is the Satyr Comma Anglewing with camouflaged underwings that match tree bark perfectly. Butterfly sizes vary from the Giant Swallowtail of southern California citrus groves that has a respectable wing-spread of nearly six inches to the smallest butterfly in the world, the Western Pigmy Blue of coastal sea bluffs, at under ¾ of an inch.

Butterflies are beloved for a number of reasons. They symbolize sunny days in the flower fields, are an inspiration for clothing and jewelry designs, and are an enticement for live-release at special ceremonies. Butterflies are also lured to our gardens, exhibited at our zoos and museums, photographed with our new digital devises, and have become field trip destinations, notably the Monarch clusters within the coastal groves of California and the mountaintop sanctuaries in central Mexico. These iconic insects are part of our culture, but their influence in the wild world is critical. Butterflies and moths are some of the planet's primary pollinators and are major participants in the food web. Their little bodies, especially the caterpillars, are important nutritional resources for birds, lizards, and a multitude of other insectivorous animals.

Southern California has a great variety of butterflies and that is a reflection of the range of habitats available to them. The ecological diversity of coastal plains, foothills, mountains, and deserts, with riparian zones in each habitat, allows for a diverse assortment of butterflies. Certain butterflies flourish in specific habitats; others are generalists and can be found almost anywhere, but a habitat without the food plant for the caterpillar will not sustain an adult population. As you scan through the book you will become familiar with the habitats and what part of the season the species fly. Melding the location of the caterpillar's food plant with the time of the year will give you a better chance of seeing a particular butterfly. This book also includes two moth species in order to point out the differences between a butterfly and a moth.

If you enjoy this *Color & Learn* book, try one of the others in this series that leads to greater understanding of the natural world. Your purchase of this book helps support the San Diego Natural History Museum. All author royalties are assigned to the Museum.

ACKNOWLEDGMENTS

This book could not have been published during the Covid-19 pandemic had it not been for the generosity of the following funders:

MAJOR FUNDERS:

Ellen Bevier

William and Terri Buchanan

Gregory and Anne Bullard

Anita Busquets and William Ladd

California Chaparral Institute

John DeBeer and Mona Baumgartel

Glenn Dunham

Donald Fosket

Dr. Bob and Linda Gordon

David H. Kaplan

Fred Kramer

Anne and Andy McCammon

Mary M. Yang, PhD

OTHER CONTRIBUTORS:

Richard T. Campbell, Wayne Cherry, Rose and Jon Cooper,
Mary Duffy, Robin Hampton, Jennifer Haslam, Mike Hughes,
Pauline K. Jimenez, Robert MacDonald, Ric Matthews,
Dale Noonkester, Jim Parker and Susan Meckel, Emily Pittman,
Jane and John Ploetz, Flavio Santoyo, Susan Stiver, Brian Swanson,
Nancy and George Varga, Jim and Terri Varnell

CATERPILLAR FOOD PLANTS: Southern California

	Common Name	Scientific Name	Food Plant / Family
1	Anise Swallowtail	*Papilio zelicaon*	Sweet Fennel (*Foeniculum vulgare*) and other members of the carrot family [Apiaceae]
2	Giant Swallowtail	*Papilio cresphontes*	Citrus Family [Rutaceae]
3	Pale Swallowtail	*Papilio eurymedon*	Buckthorns [Rhamnaceae], wild roses [Rosaceae], and other chaparral shrubs
4	Western Tiger Swallowtail	*Papilio rutulus*	Willows, cottonwoods [Salicaceae], sycamore [Platanaceae], and other riparian trees
5	Checkered White	*Pontia protodice*	Mustard Family [Brassicaceae]
6	Sara Orangetip	*Anthocharis sara*	Mustard Family [Brassicaceae]
7	Orange Sulphur	*Colias eurytheme*	Alfalfa (*Medicago sativa*) and other legumes [Fabaceae]
8	California Dogface	*Colias eurydice*	False indigo (*Amorpha californica*) [Fabaceae]
9	Cloudless Sulphur	*Phoebis sennae*	Cassia (*Senna*) [Fabaceae]
10	Hermes Copper	*Lycaena hermes*	Spiny redberry (*Rhamnus crocea*) [Rhamnaceae]
11	Great Purple Hairstreak	*Atlides halesus*	Mistletoe Family [Viscaceae]
12	Gray Hairstreak	*Strymon melinus*	Mallows [Malvaceae] and many other plant families
13	Juniper Hairstreak	*Callophrys gryneus*	Cypress Family [Cupressaceae]
14	Sonoran Blue	*Philotes sonorensis*	Live-Forevers (*Dudleya*) [Crassulaceae]
15	Western Pygmy Blue	*Brephidium exilis*	Saltbush (*Atriplex*) [Chenopodiaceae] and related plants
16	Melissa Blue	*Lycaeides melissa*	Lupines (*Lupinus*) and other legumes [Fabaceae]
17	Acmon Blue	*Plebejus acmon*	Legumes [Fabaceae] and buckwheats (*Eriogonum*) [Polygonaceae]
18	Silvery Blue	*Glaucopsyche lygdamus*	Deerweed (*Acmispon glaber*) and other legumes [Fabaceae]
19	Marine Blue	*Leptotes marina*	Legumes [Fabaceae] and Plumbago [Plumbaginaceae]
20	Arrowhead Blue	*Glaucopsyche piasus*	Lupines (*Lupinus*) [Fabaceae]
21	Mormon Metalmark	*Apodemia mormo*	Buckwheats (*Eriogonum*) [Polygonaceae]
22	Coronis Fritillary	*Speyeria coronis*	Violets (*Viola*) [Violaceae]
23	Gulf Fritillary	*Agraulis vanillae*	Passion Vines (*Passiflora*) [Passifloraceae]
24	Variable Checkerspot	*Euphydryas chalcedona*	Monkeyflower [Phrymaceae] and related plants
25	Quino Checkerspot	*Euphydryas editha quino*	California plantain (*Plantago erecta*) [Plantaginaceae] and related plants
26	Satyr Comma Anglewing	*Polygonia satyrus*	Nettle (*Urtica dioica*) [Urticaceae]
27	Mourning Cloak	*Nymphalis antiopa*	Willows, cottonwoods [Salicaceae], elms [Ulmaceae], and other plants
28	American Lady	*Vanessa virginiensis*	Cudweed (*Pseudognaphalium*) [Asteraceae]
29	Painted Lady	*Vanessa cardui*	Many different plant families
30	West Coast Lady	*Vanessa annabella*	Cheeseweed (*Malva parviflora*) and other mallows [Malvaceae]
31	Common Buckeye	*Junonia coenia*	Plantains, snapdragons [Plantaginaceae], monkeyflowers [Phrymaceae], and other related plants
32	Lorquin's Admiral	*Limenitis lorquini*	Willows [Salicaceae] and related plants
33	California Sister	*Adelpha bredowii*	Oaks [Fagaceae]
34-37	Monarch	*Danaus plexippus*	Milkweeds [Apocynaceae]
37	Queen	*Danaus gilippus*	Milkweeds [Apocynaceae]
38	Fiery Skipper	*Hylephila phyleus*	Grasses [Poaceae]
38	Juba Skipper	*Hesperia juba*	Grasses [Poaceae]
39	Ceanothus Silk Moth	*Hyalophora euryalus*	California Lilacs (*Ceanothus*) [Rhamnaceae] and other woody shrubs
40	White-Lined Sphinx Moth	*Hyles lineata*	Evening-Primrose Family [Onagraceae] and other plant families

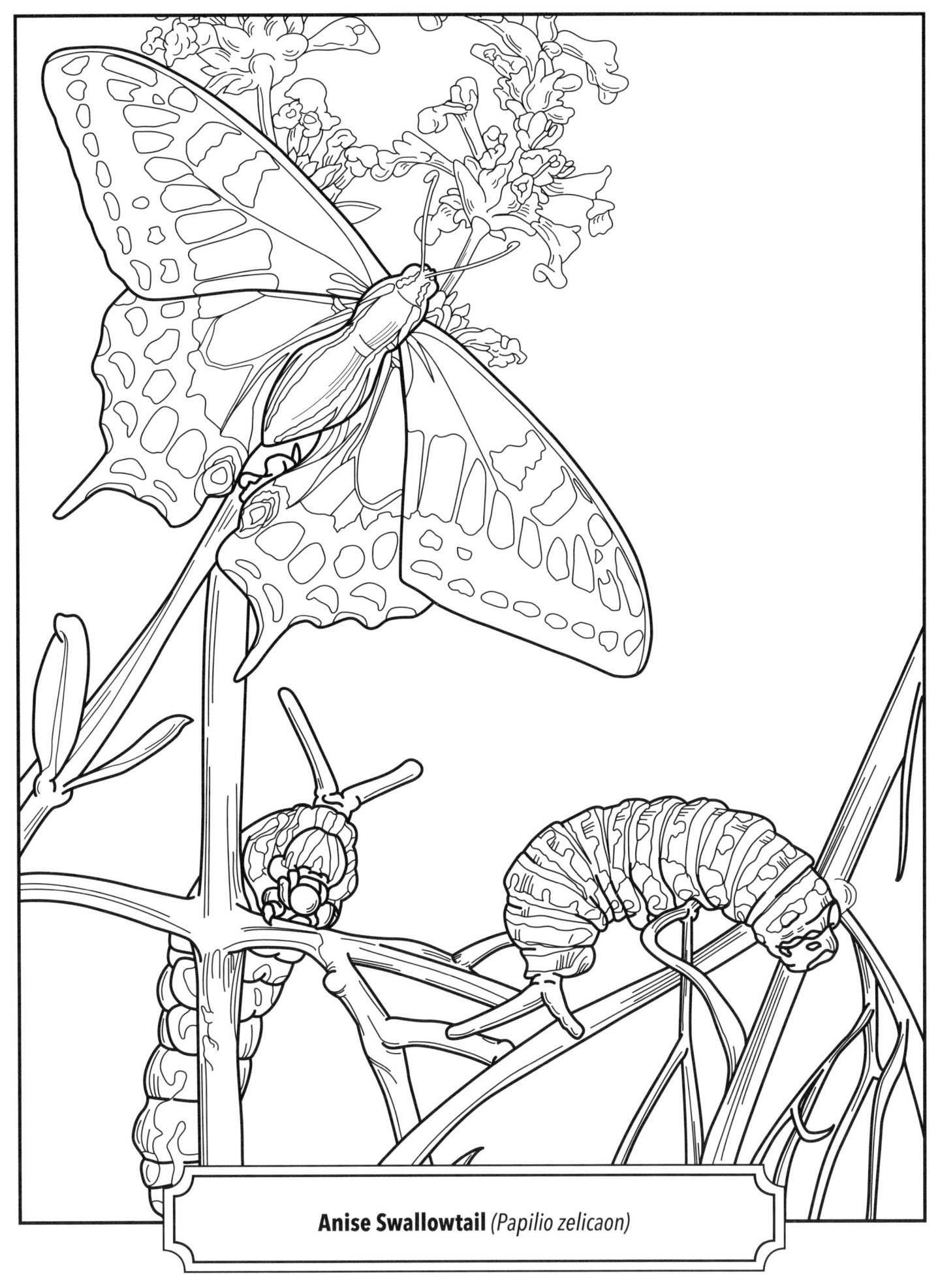

Anise Swallowtail *(Papilio zelicaon)*

Giant Swallowtail *(Papilio cresphontes)*

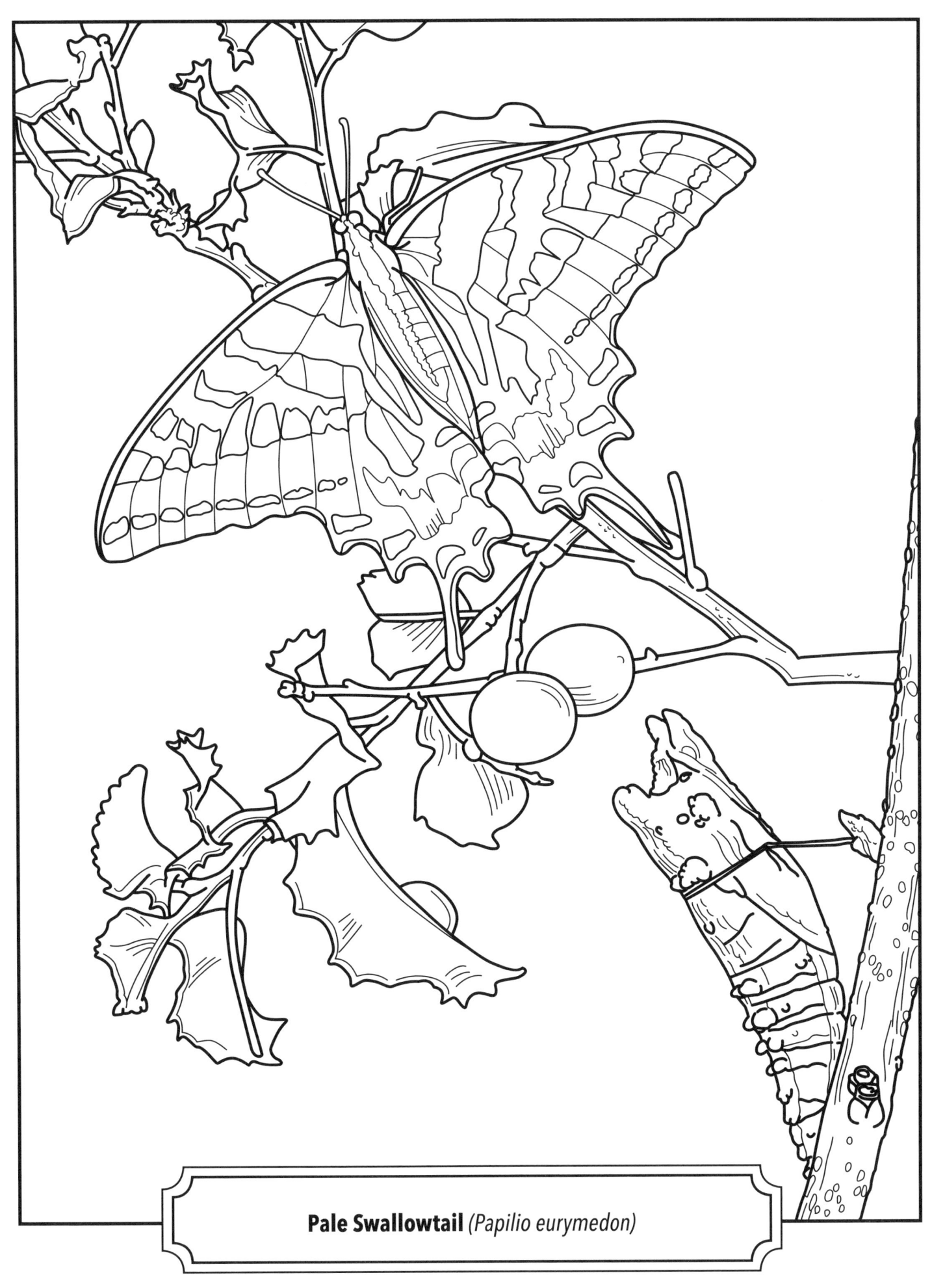

Pale Swallowtail *(Papilio eurymedon)*

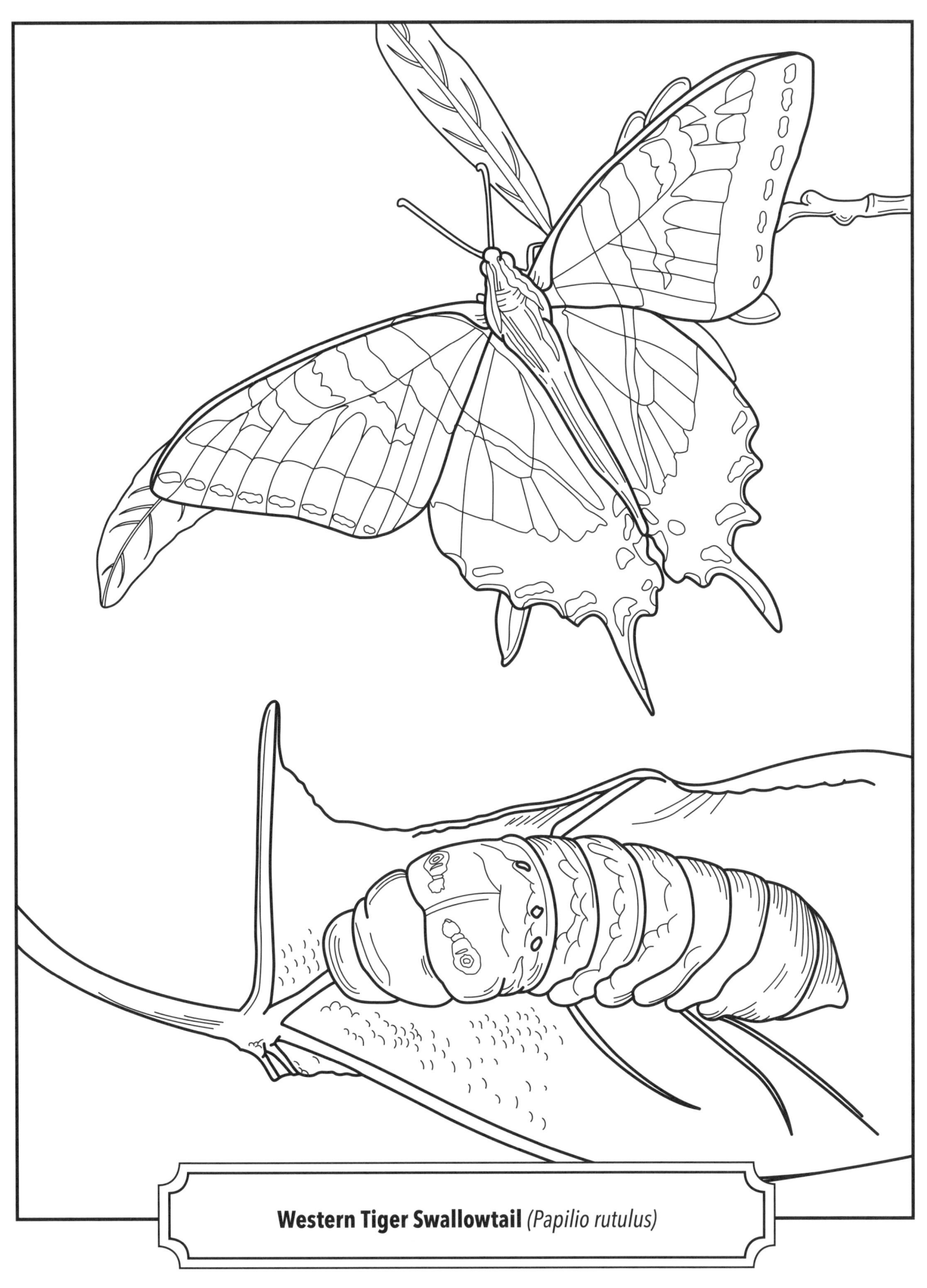

Western Tiger Swallowtail *(Papilio rutulus)*

Checkered White *(Pontia protodice)*

Sara Orangetip *(Anthocharis sara)*

Orange Sulphur (Colias eurytheme)

California Dogface (*Colias eurydice*)

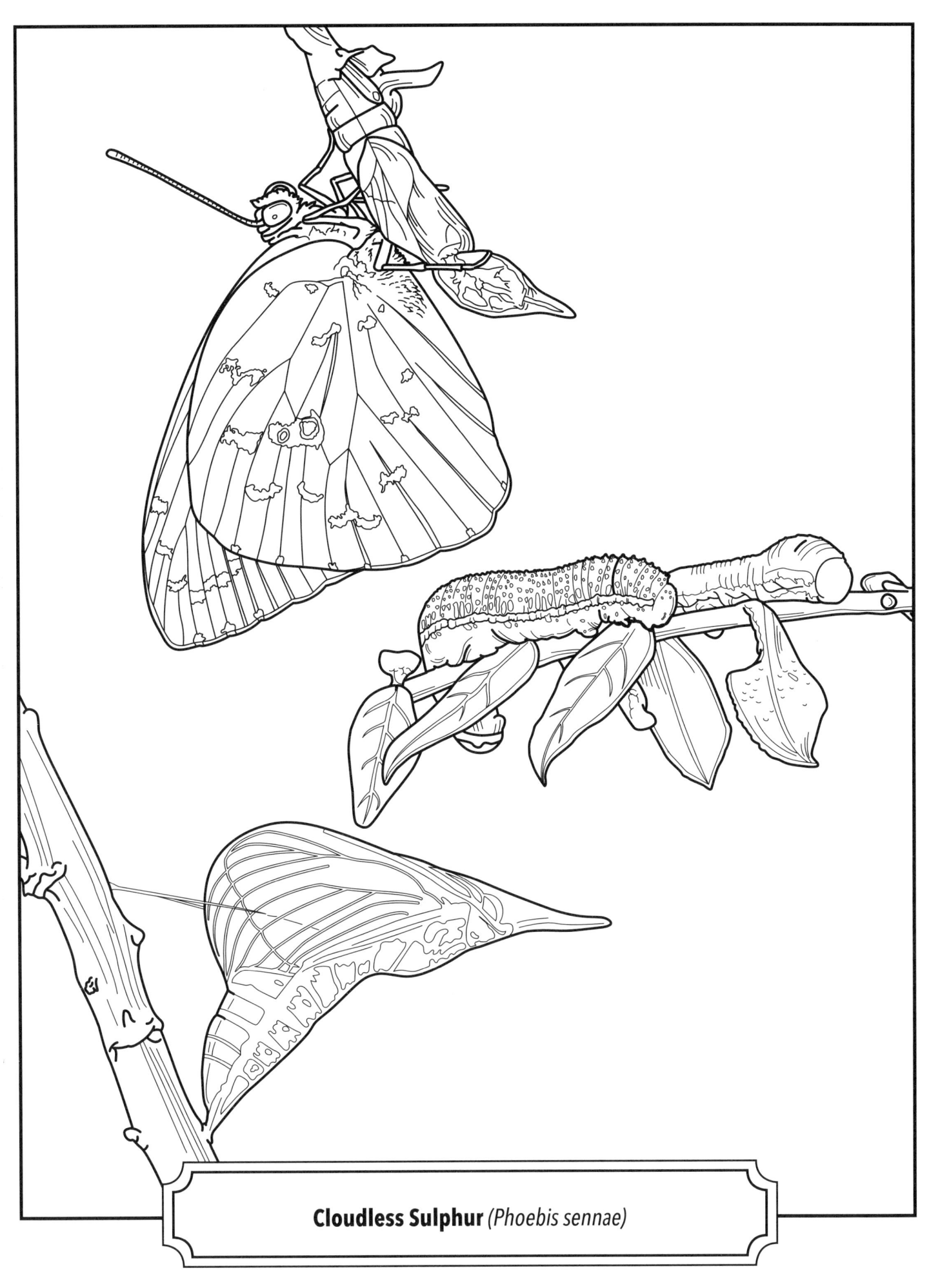

Cloudless Sulphur *(Phoebis sennae)*

Hermes Copper (Lycaena hermes)

Great Purple Hairstreak *(Atlides halesus)*

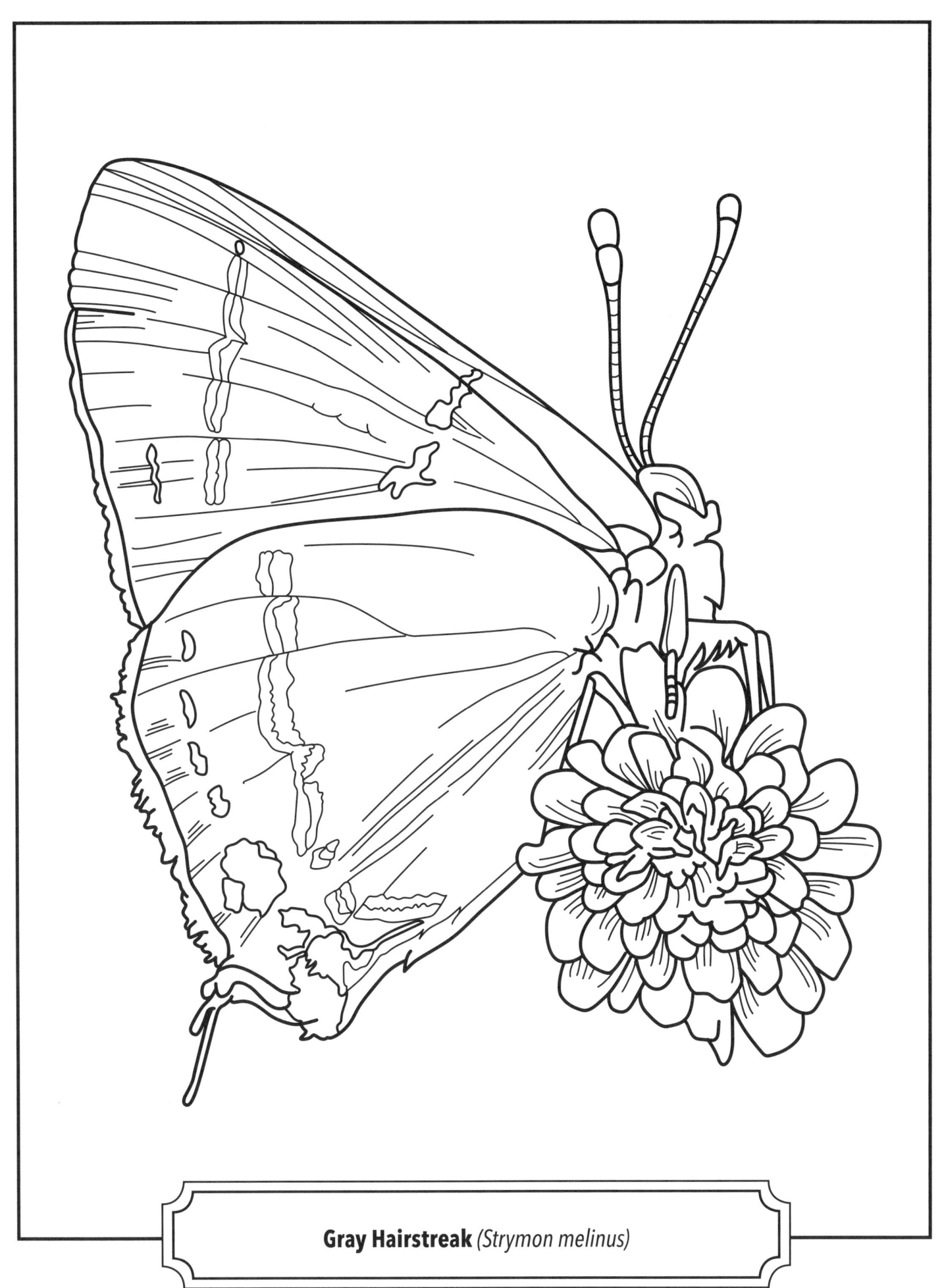

Gray Hairstreak (*Strymon melinus*)

Juniper Hairstreak *(Callophrys gryneus)*

Sonoran Blue *(Philotes sonorensis)*

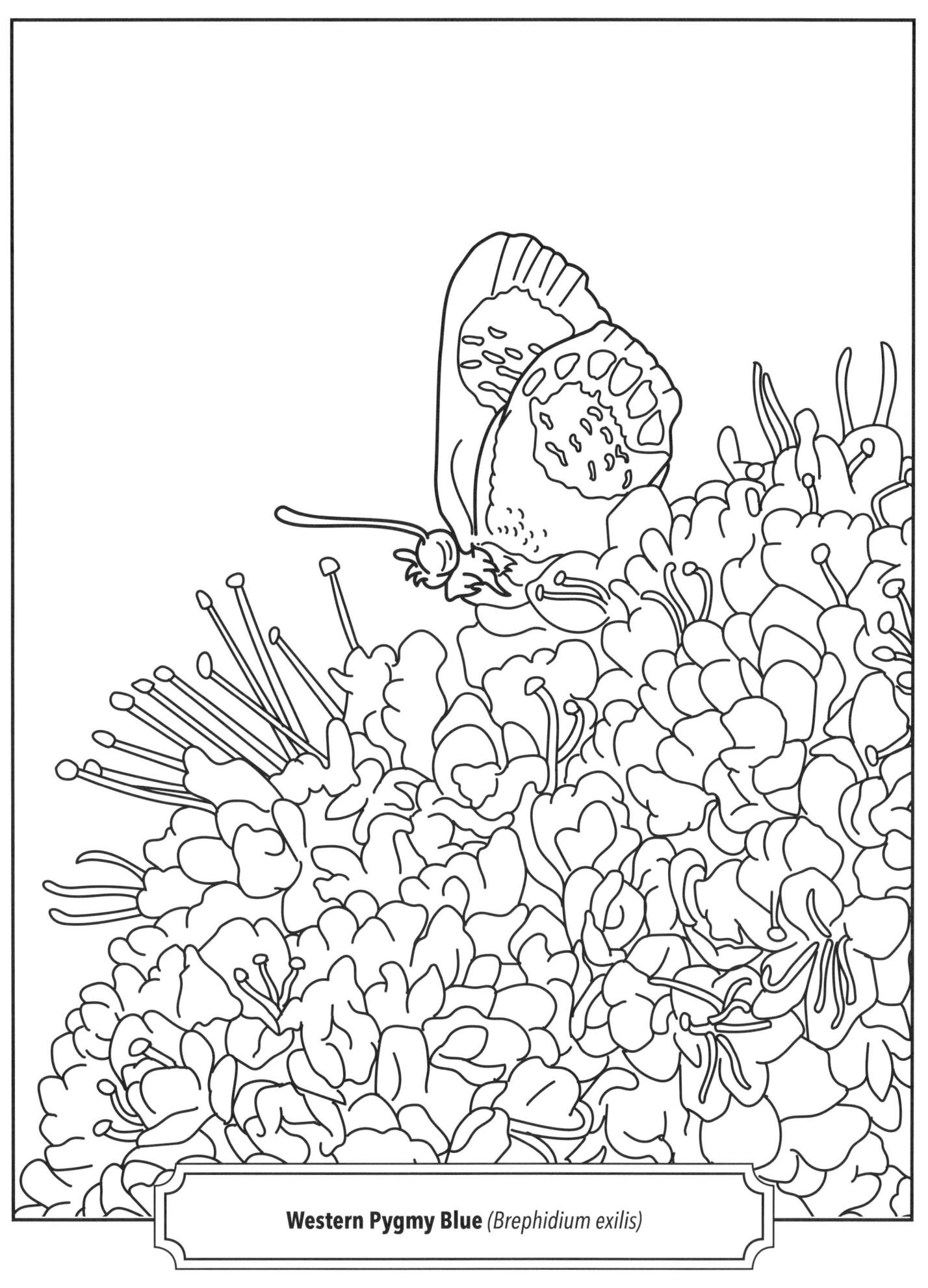

Western Pygmy Blue *(Brephidium exilis)*

Melissa Blue *(Lycaeides melissa)*

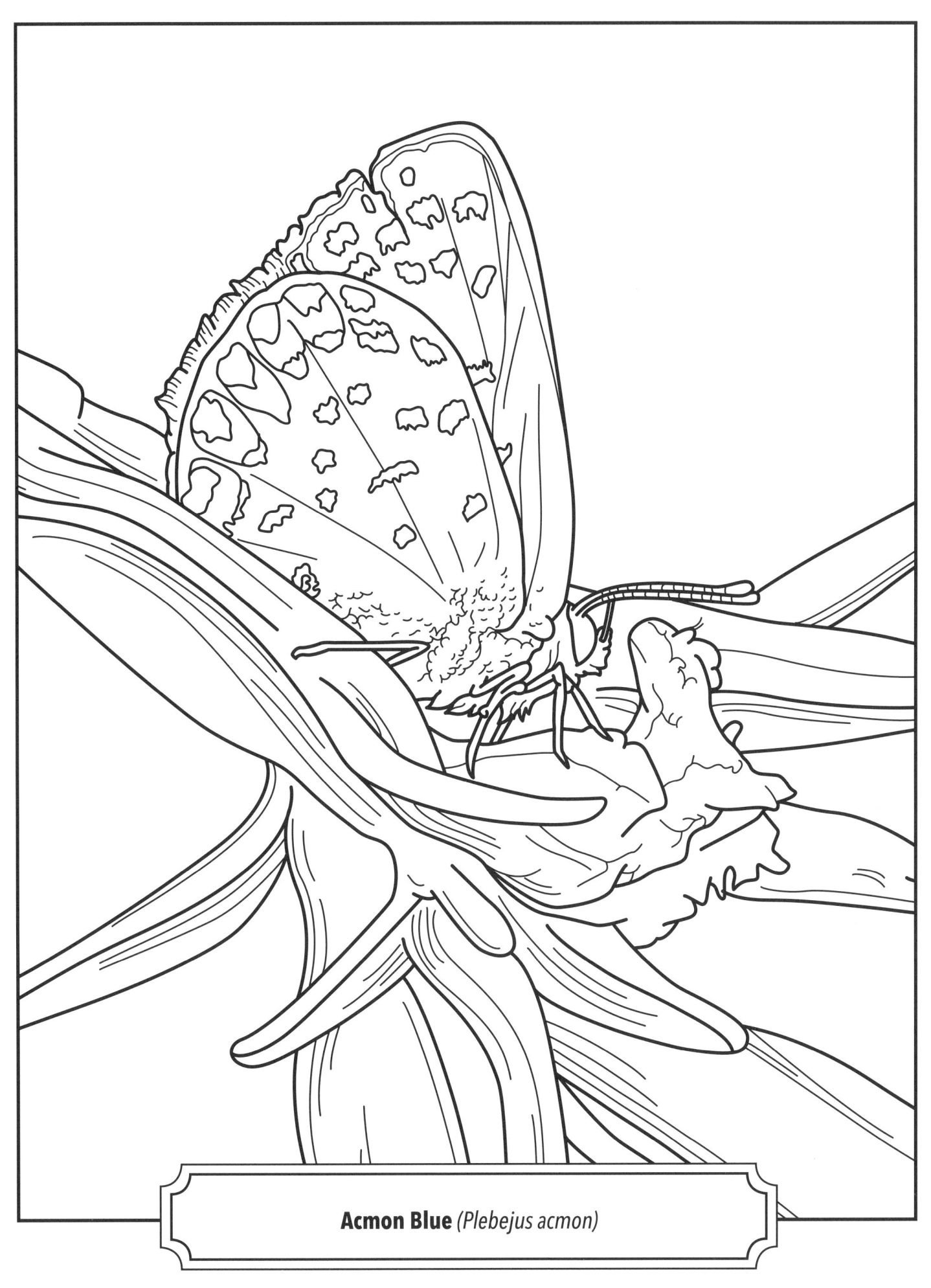

Acmon Blue *(Plebejus acmon)*

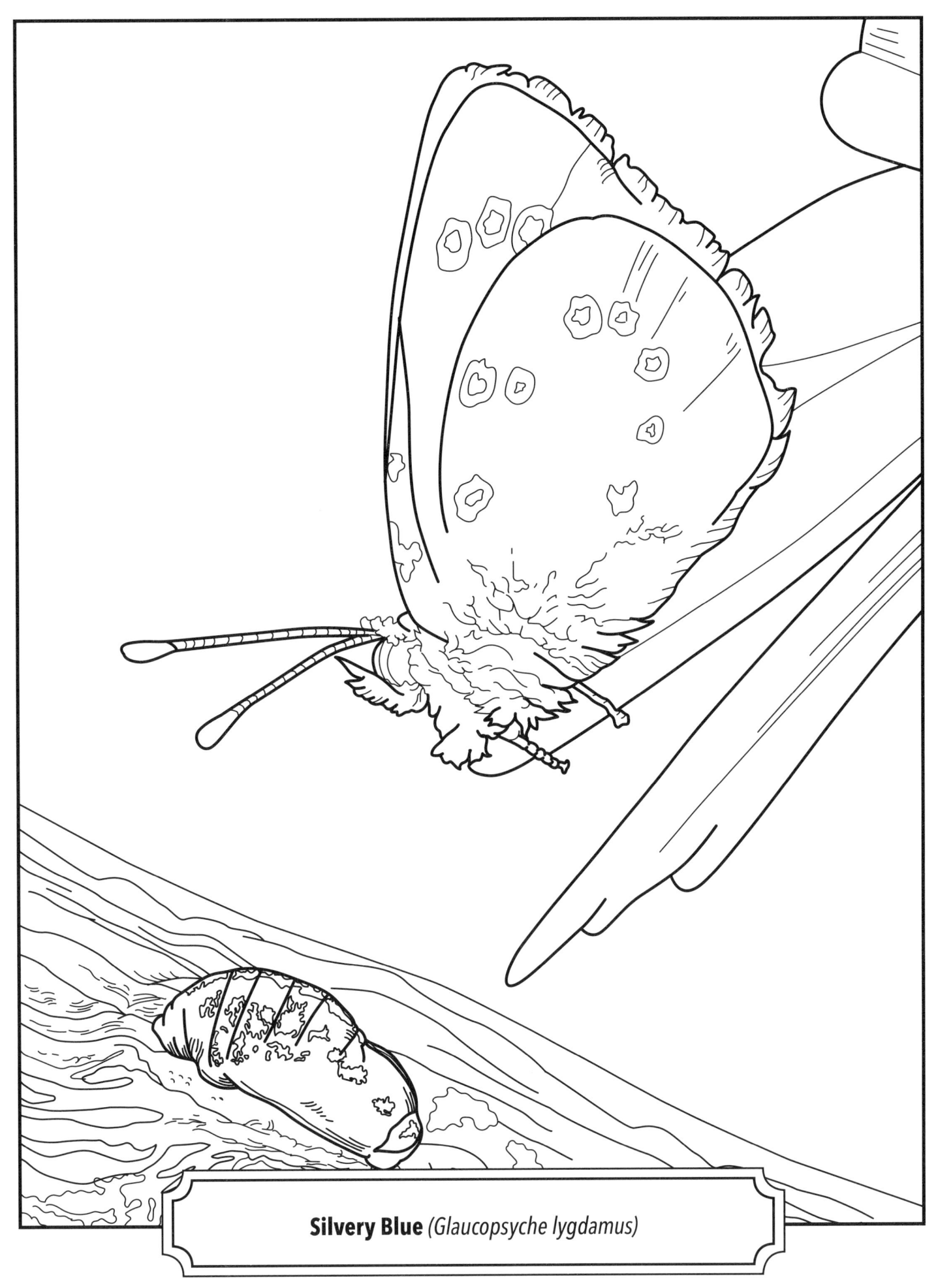

Silvery Blue (*Glaucopsyche lygdamus*)

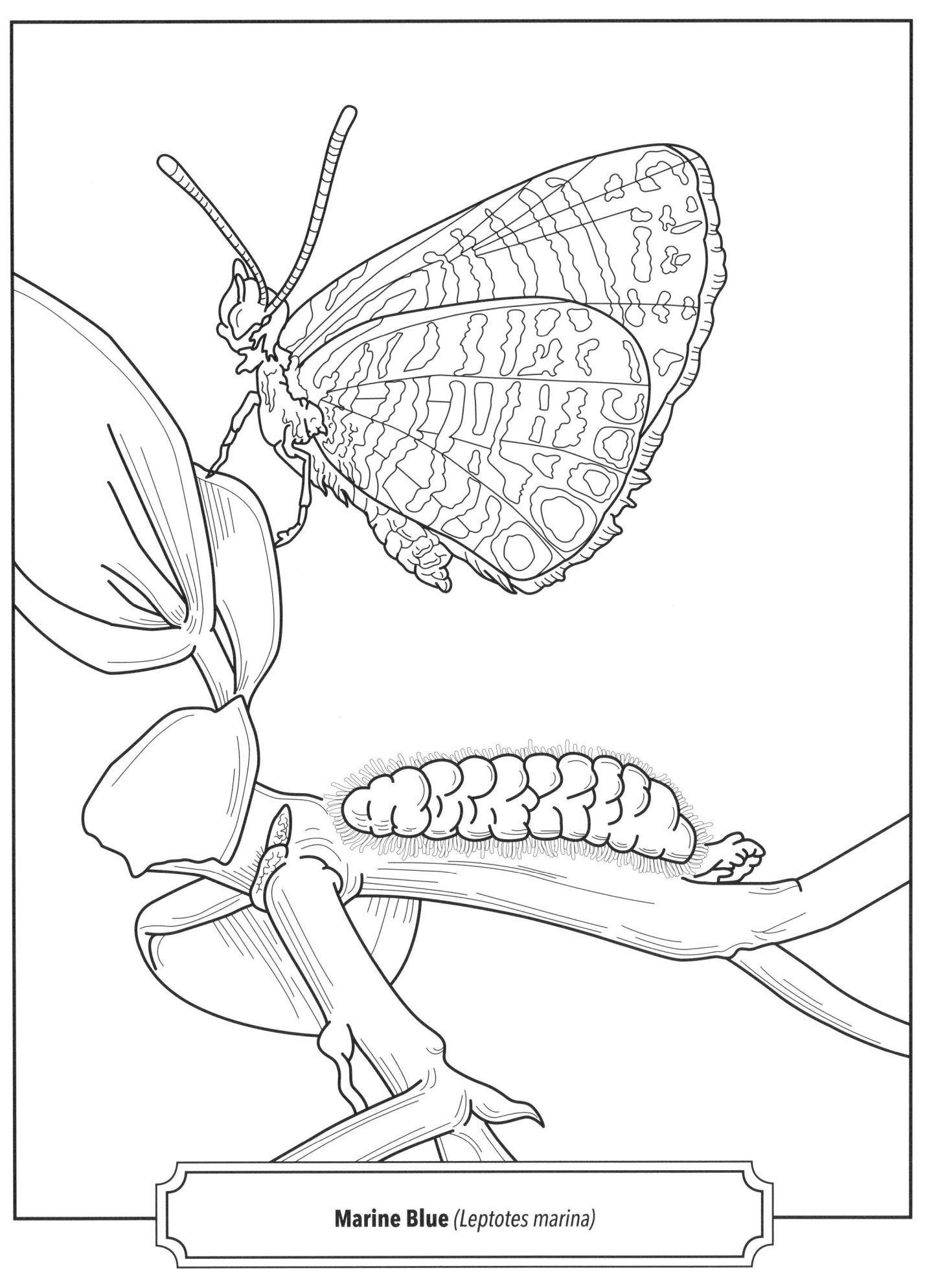

Marine Blue *(Leptotes marina)*

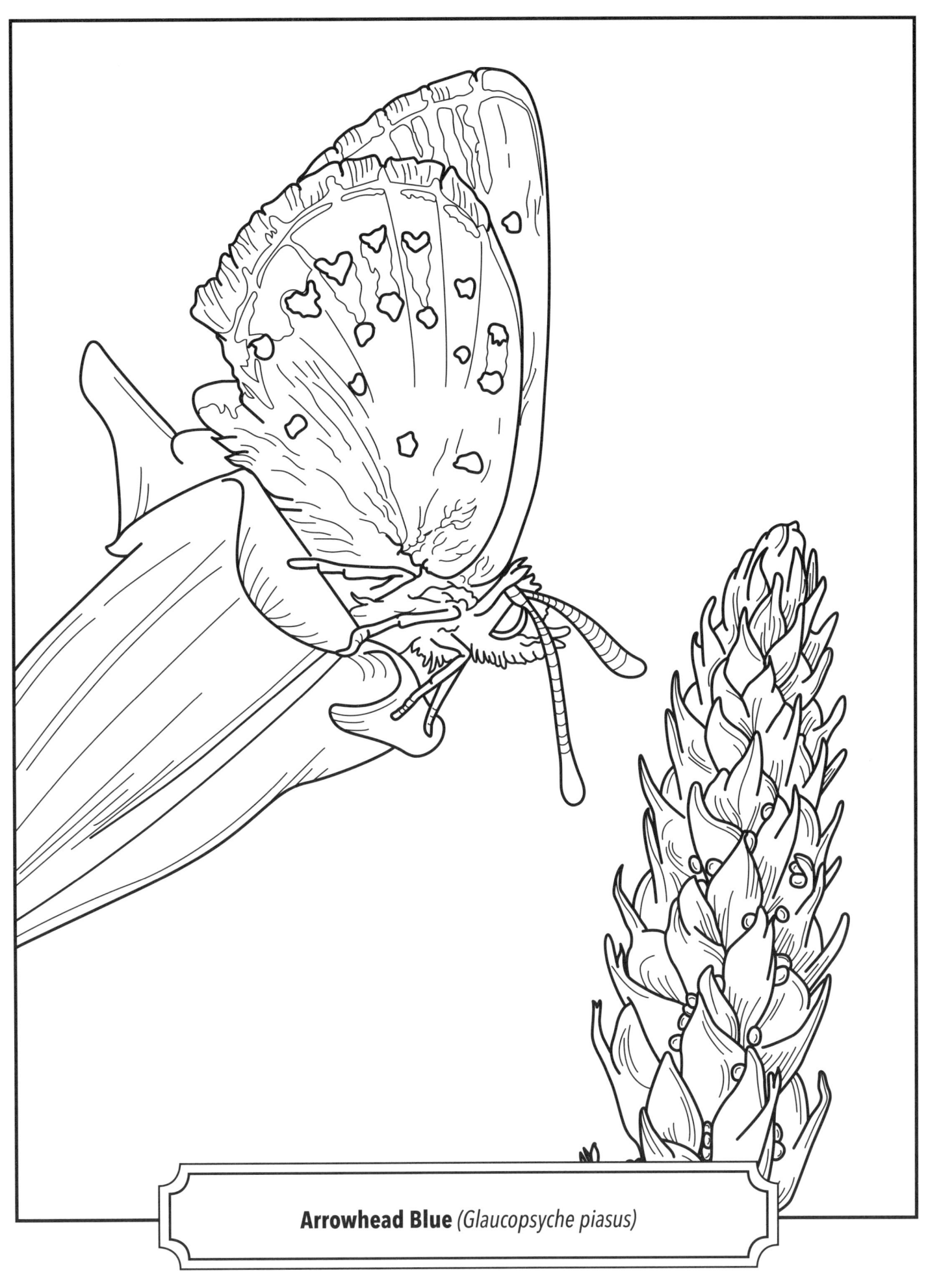

Arrowhead Blue *(Glaucopsyche piasus)*

Mormon Metalmark *(Apodemia mormo)*

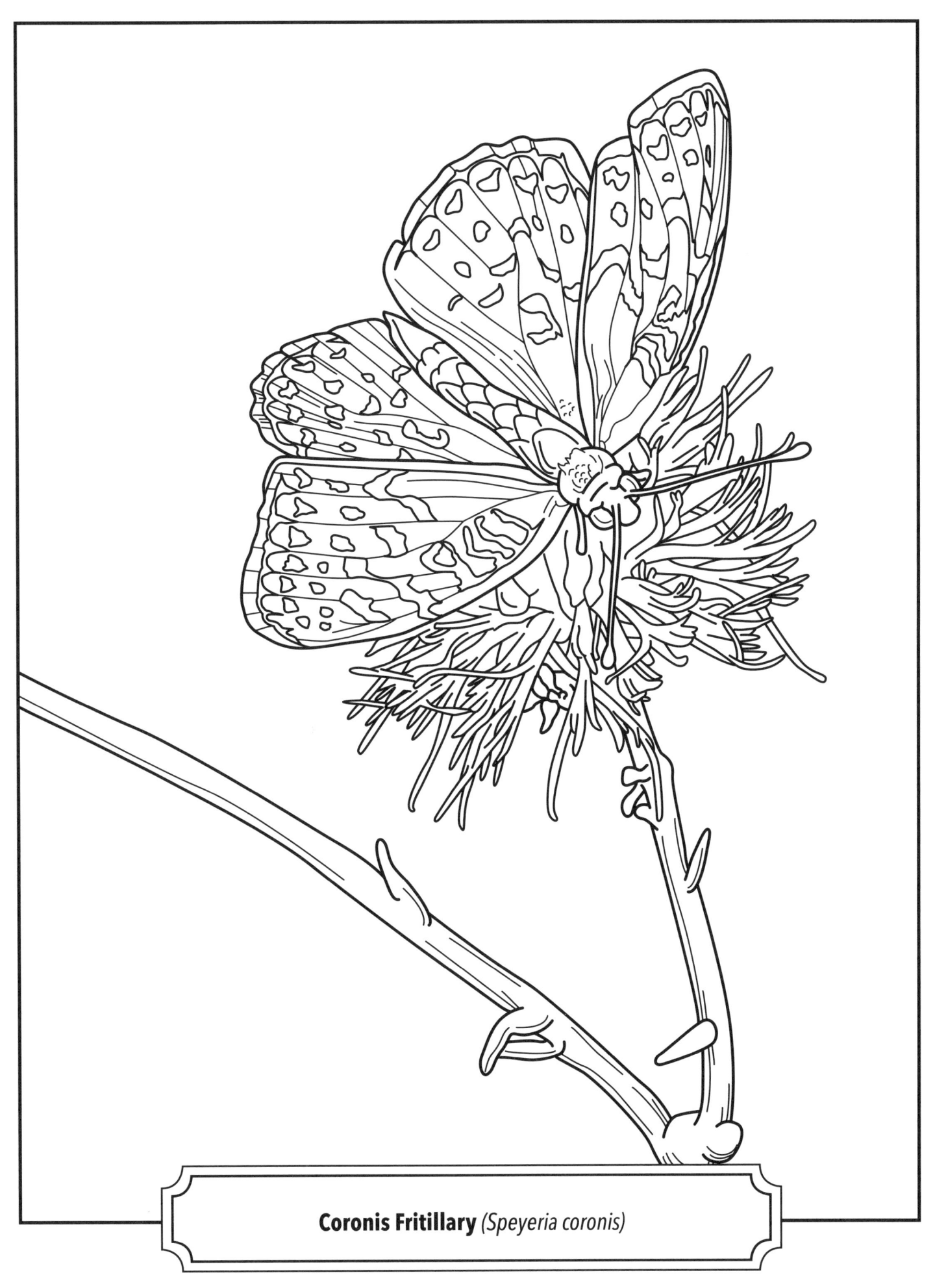

Coronis Fritillary *(Speyeria coronis)*

Gulf Fritillary *(Agraulis vanillae)*

Variable Checkerspot *(Euphydryas chalcedona)*

Quino Checkerspot *(Euphydryas editha quino)*

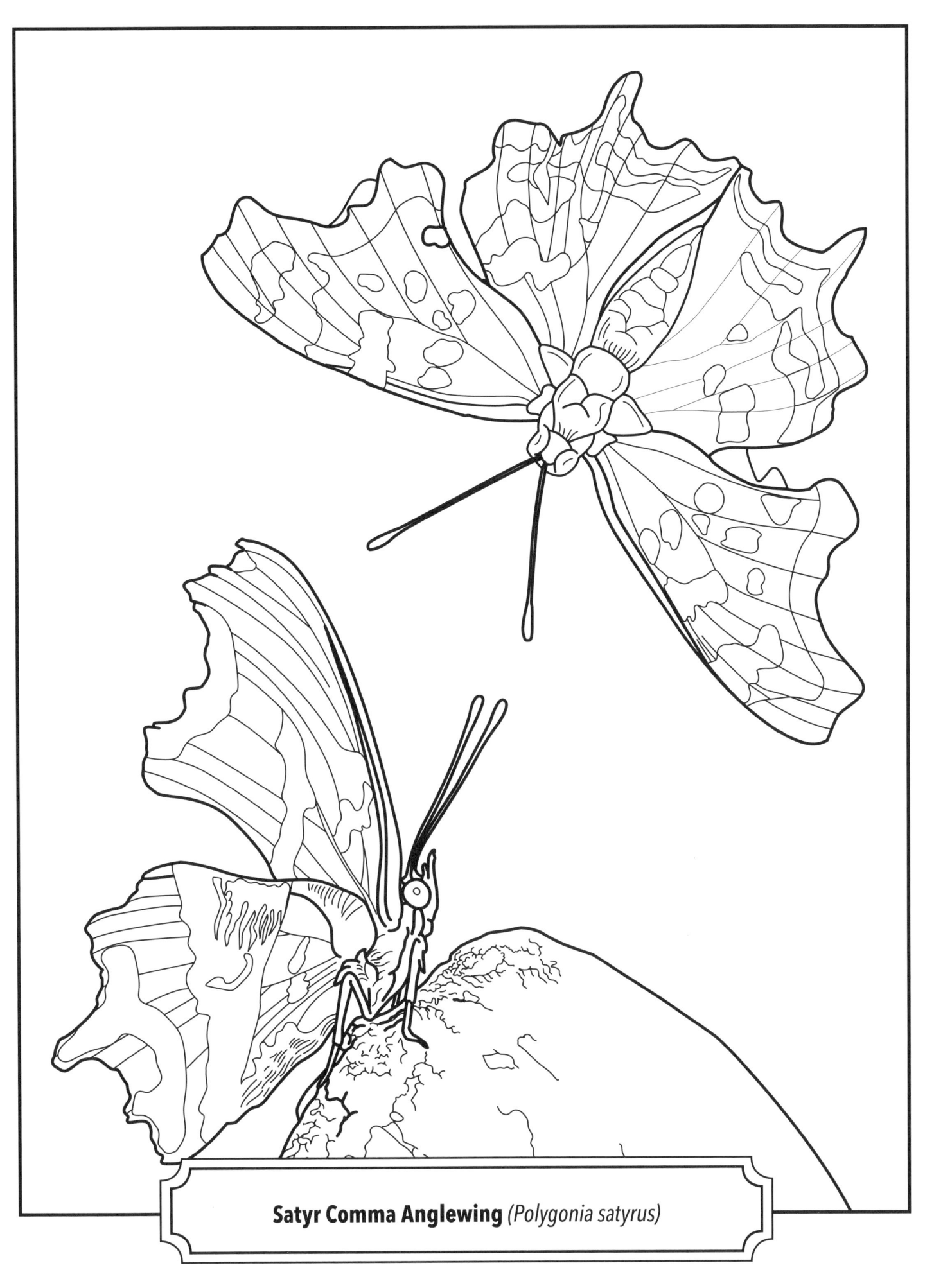
Satyr Comma Anglewing *(Polygonia satyrus)*

Mourning Cloak (*Nymphalis antiopa*)

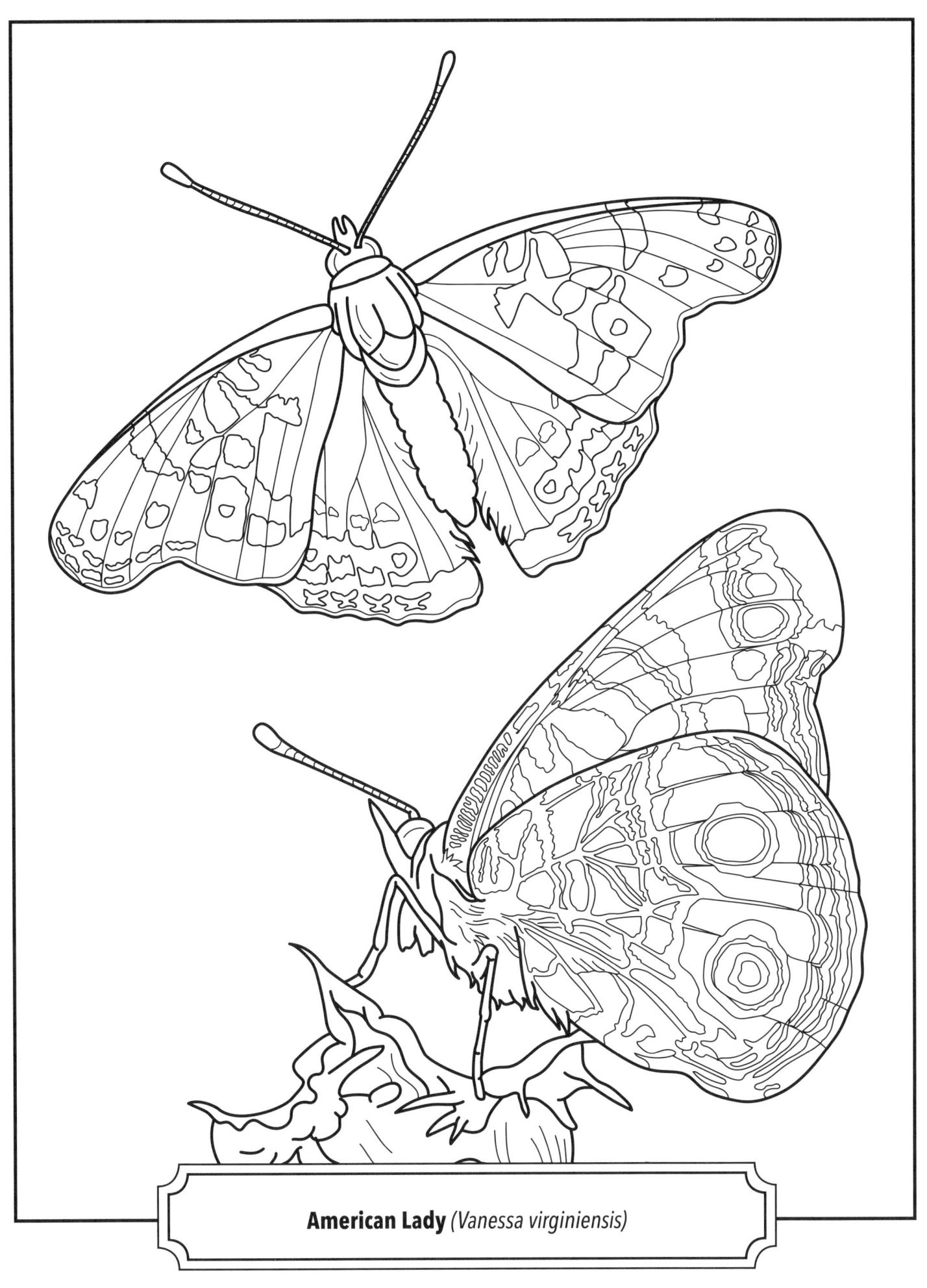

American Lady *(Vanessa virginiensis)*

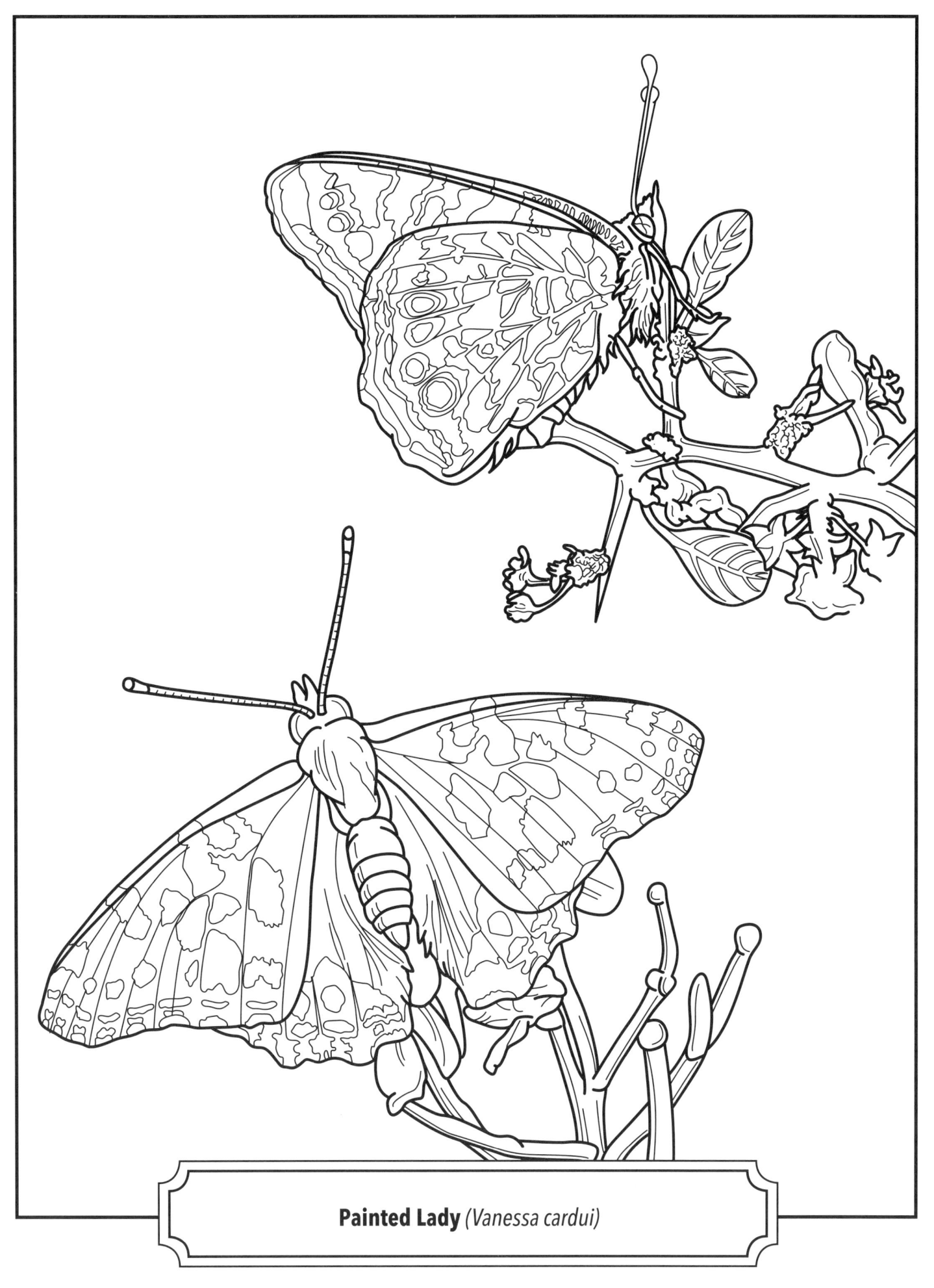

Painted Lady *(Vanessa cardui)*

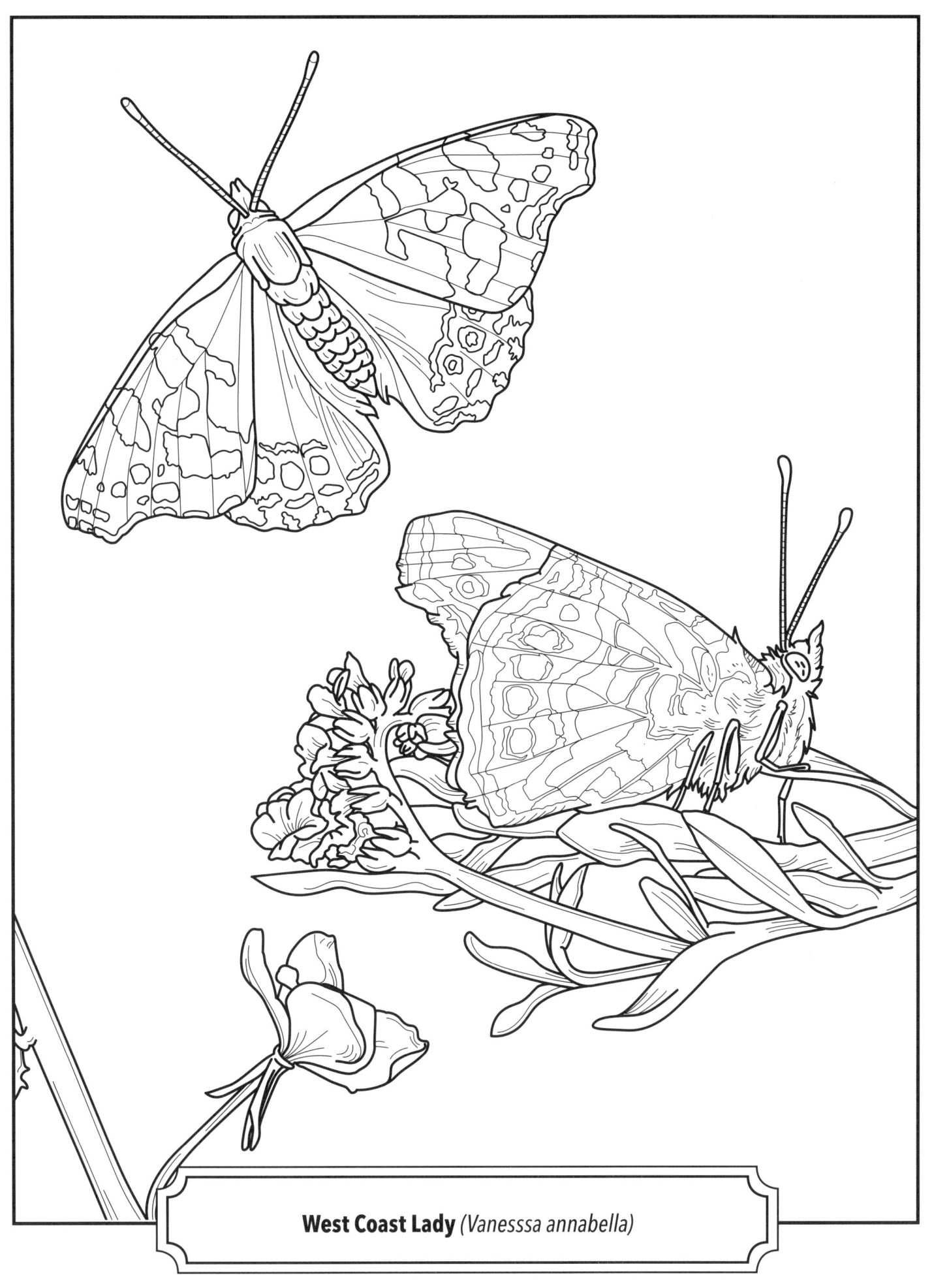

West Coast Lady *(Vanesssa annabella)*

Common Buckeye *(Junonia coenia)*

Lorquin's Admiral *(Limenitis lorquini)*

California Sister (Adelpha bredowii)

Monarch *(Danaus plexippus)*

Monarch Caterpillar *(Larva)* **and Chrysalis** *(Pupa)*

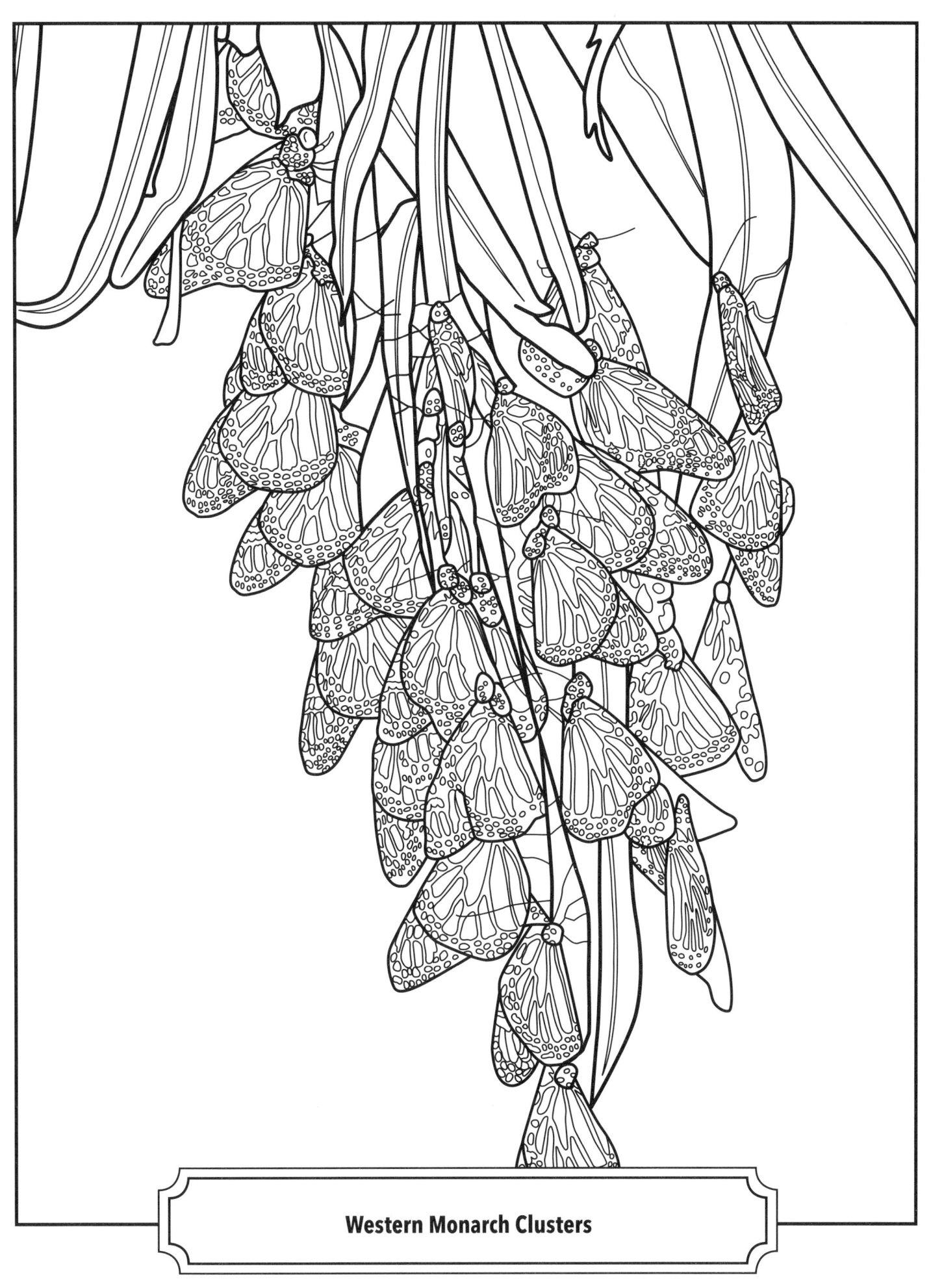

Western Monarch Clusters

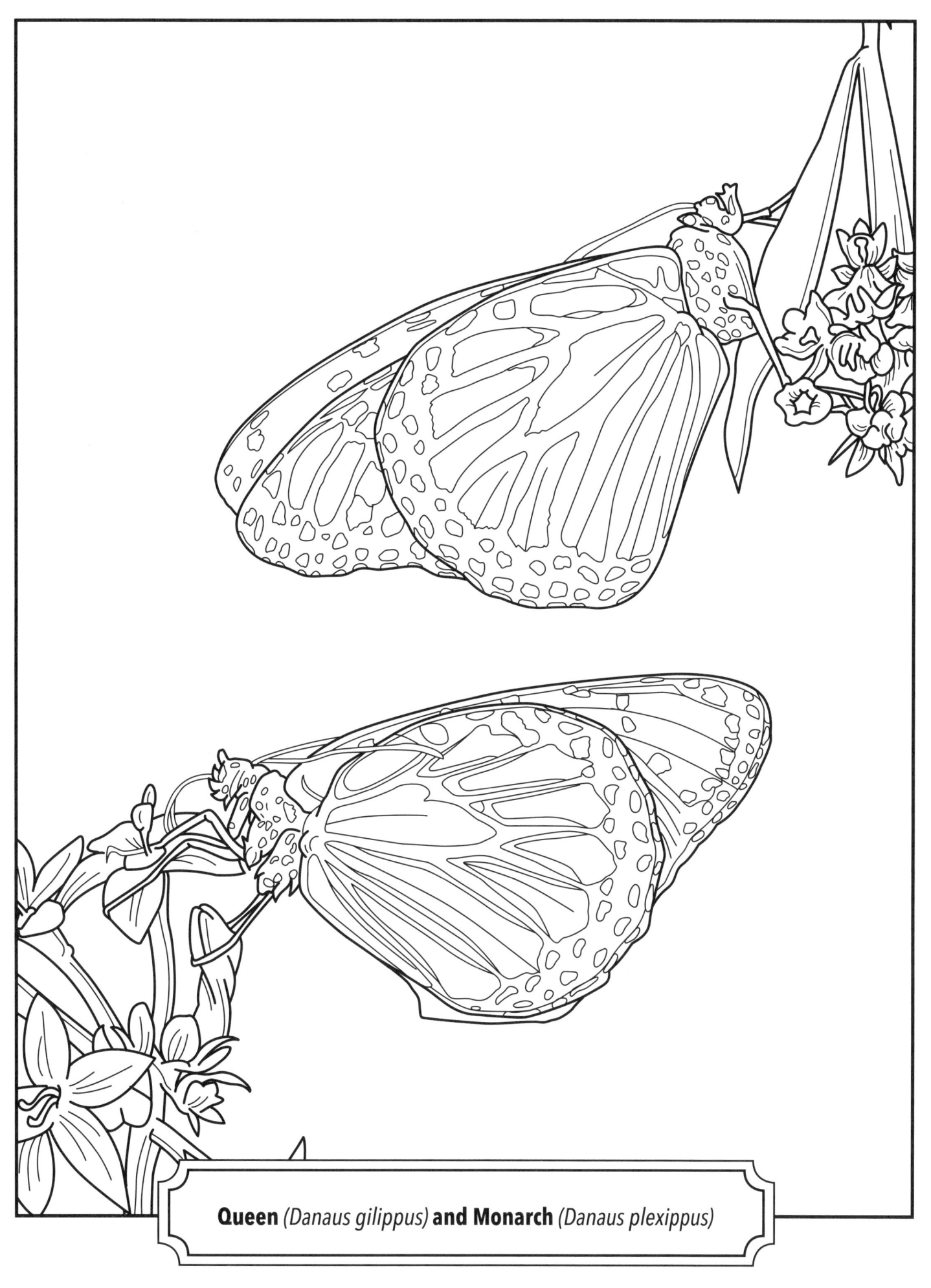

Queen *(Danaus gilippus)* **and Monarch** *(Danaus plexippus)*

Skippers: [**Fiery** (*Hylephila phyleus*), **Juba** (*Hesperia juba*)]

Ceanothus Silk Moth (*Hyalophora euryalus*)

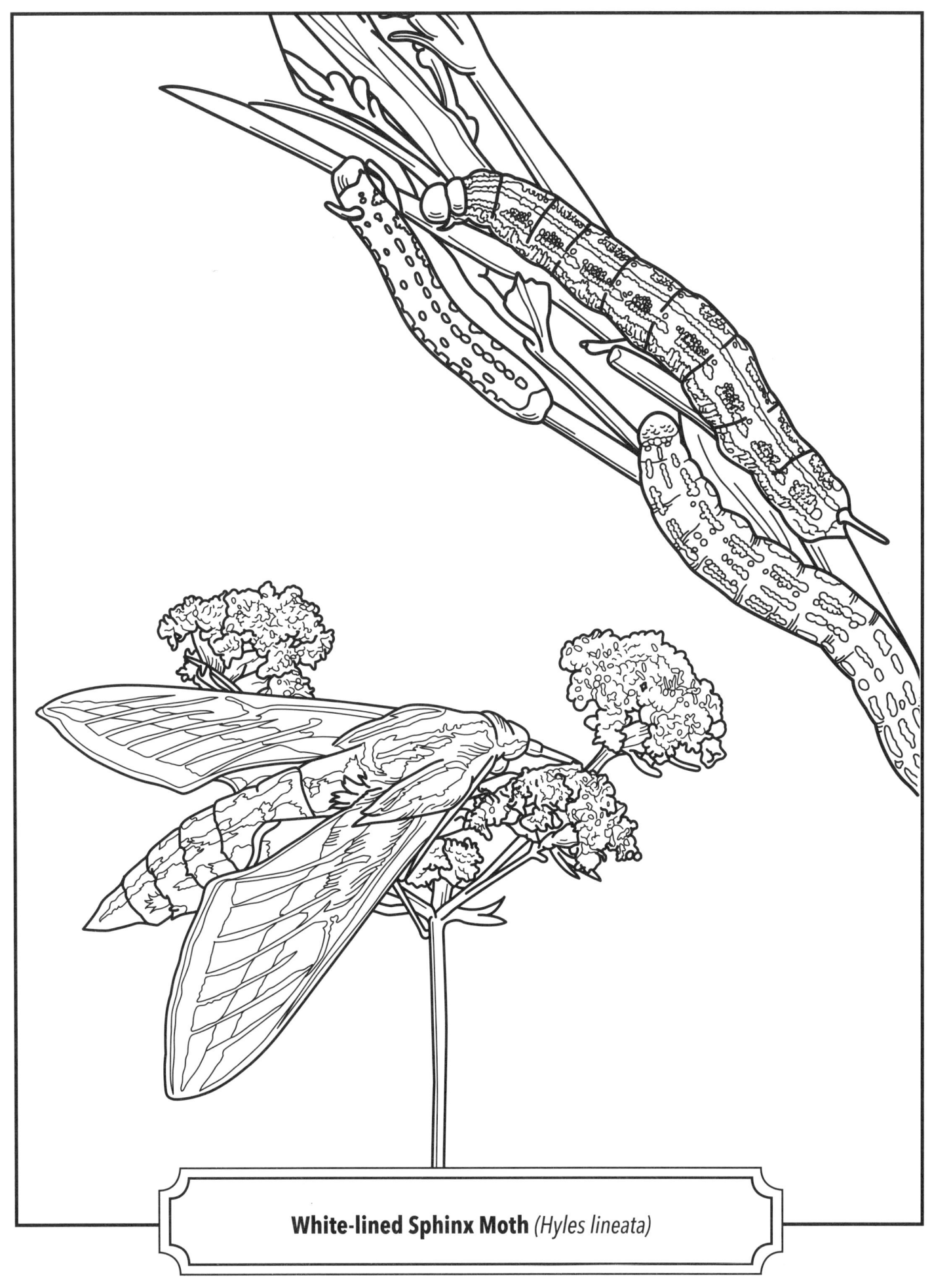

White-lined Sphinx Moth *(Hyles lineata)*

1) Anise Swallowtail *(Papilio zelicaon)*   2⅝ – 3"

The Anise Swallowtail butterfly may be the most extensively seen swallowtail in southern California. It extends into southern Canada and is common in the western United States except in desert areas. Like most (but not all) swallowtail butterflies it has a swallow-like tail on each hindwing. The naked green caterpillar has no hairs or filaments and has bands of black and pale yellow spots. All swallowtail caterpillars, if disturbed, extend a stinky, orange bifurcated protrusion from the back of their heads to allegedly deter predators. The forked projection is called an osmeterium. The food plant for the caterpillar includes members of the carrot family with fennel being a favorite. The chrysalis is held upright with a necklace of silk and ranges in color from bark-brown to leaf-green which suggests a camouflage strategy.

2) Giant Swallowtail *(Papilio cresphontes)*   3⅜ – 5½"

The Giant Swallowtail is also known as the Citrus Swallowtail because the caterpillar munches on members of the citrus family, earning the moniker "orange dog." This large butterfly was not well received by southern California citrus growers when it fluttered in from the southeastern United States in the 1960s. Even with the caterpillar's poor food choice, it is not considered a major agricultural pest. For protection the caterpillar resembles bird dung to avoid being eaten. Also, it will extend a scented snake tongue-like organ (the osmeterium) from behind the head when threatened. A dark brown topside with yellow bands across the forewings and margins of the hindwings distinguishes it from other swallowtails. It is the largest butterfly in California with over a five-inch wingspan, while its cousin, the Birdwing Butterfly of the tropics, is the largest in the world with a wingspan over ten-inches.

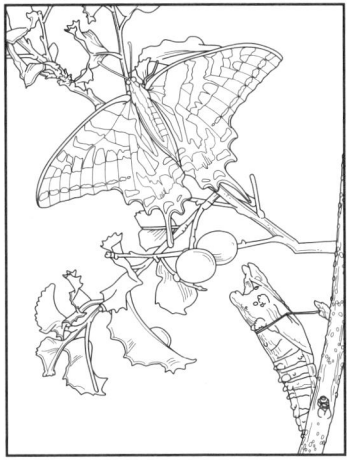

3) Pale Swallowtail *(Papilio eurymedon)*   2¾ – 3⅞"

The Pale Swallowtail butterfly is without yellow on its topside where it has a background color of cream with tiger-stripes and hindwing margins of black. Like most swallowtail butterflies, it has a tail from each of the hindwings. From shoreline to timberline it is a fast and erratic flier along chaparral canyons and foothills. It will occasionally visit riparian areas where it flies in spring and summer with the Western Tiger Swallowtail, which it resembles. The caterpillar has fake eyespots like the Western Tiger Swallowtail and has a fleshy-forked organ (the osmeterium) that protrudes from behind its head when alarmed. The caterpillars feed on certain members of the buckthorn and rose family. The chrysalis is usually attached to a twig or leaf and held upright with a silk necklace spun from the silk gland of the caterpillar. This special feature will identify any swallowtail pupa.

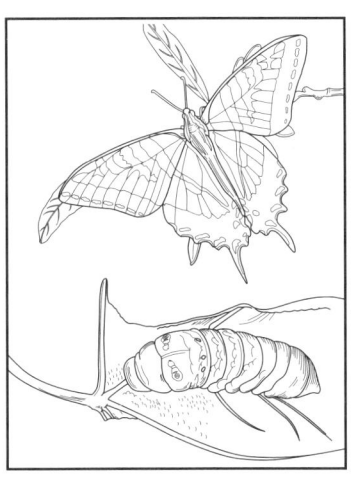

4) Western Tiger Swallowtail *(Papilio rutulus)*   3 – 3¾"

The Western Tiger Swallowtail can easily be found along non-desert riparian corridors in most of southern California. It is called a swallowtail because it has tails on the hindwings like the bird. This yellow butterfly with black "tiger" stripes is one of the largest butterflies in California with a wingspan of over three inches. Although it prefers to fly between April and November, it can be seen along almost any wet zone all year. Common plants like sycamore, willow, cottonwood, and other streamside trees and shrubs are food for the green caterpillars. The caterpillar has fake eyes behind its head to allegedly scare away would-be predators. It also has the orange antler-like, smelly osmeterium that emerges from the same region when alarmed. All swallowtail caterpillars have this osmeterium, but not all swallowtails have tails. The dark brown, horned chrysalis is attached upright on a twig with a thin necklace of silk.

5) Checkered White *(Pontia protodice)*   1½ – 1¾"

The familiar Checkered White, more widespread from early spring into fall, may be seen every sunny day of the year in southern California. This low flying insect with rapid erratic flight is easily observed from coastal regions to deserts in weedy fields and disturbed areas. The upperside of the female forewing has heavily checkered markings while the male has faint patterns or none; both have yellow-brown veins below. The caterpillars favor the buds, flowers, fruits, and occasionally the leaves of members of the mustard family. The males patrol hilltops seeking females by using ultraviolet light rather than pheromones.

6) Sara Orangetip *(Anthocharis sara)*   1¼ – 1¾"

The Sara Orangetip butterfly signals a new season as winter wanes. It flies with rapid wing beats through early springtime meadows, canyon slopes, and foothill arroyos while the males patrol for females seeking flowers. On the topside he has an unbroken black line separating the orange tips from the white wings. The female line is open. Both sexes have yellow veins and dark marbling on the underside of the hindwing. These butterflies range throughout most of the West as far north as Alaska and south into coastal Baja California. The tiny green caterpillars eat leaves of most members of the mustard family. Older caterpillars add petals and even fruits to their diet. The chemicals and oils from these cruciferous plants remain in the adults and make these little white butterflies distasteful to many vertebrate predators

7) Orange Sulphur *(Colias eurytheme)*   1⅝ – 2⅜"

The Orange Sulphur butterfly is the most widespread sulphur in North America. Other sulphurs are without any orange coloration, so an orange hue on a yellow butterfly identifies it as an Orange Sulphur. The caterpillar munches on legumes, especially alfalfa. Another name for this insect is the Alfalfa butterfly. In cultivated areas they can be viewed by the thousands fluttering through fields of alfalfa as an agricultural pest. Other habitats include mountains, foothills, and deserts. Their range is extensive and includes southern Canada, Mexico, and all of the United States. In southern California it flies almost all months of the year. Males reflect ultraviolet light that attracts females

8) California Dogface *(Colias eurydice)*   1⅞ – 2½"

The California Dogface is found only in California and Baja California and is the official state insect of California because of the unique design found on the male. To some people, the image resembles the head of a poodle dog on each of his upper forewings. The female is all yellow without the dog image and both sexes have forewings pointier than the other sulphur butterflies. Flight time is spring to fall in mountains and foothills. It may be seen in chaparral and woodland openings, especially along watercourses, where California false indigo, the favorite caterpillar food plant, grows. It is sometimes called the flying pansy because of the colors and patterns on the upper forewings of the male.

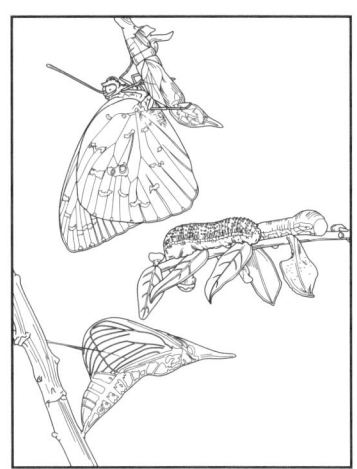

9) Cloudless Sulphur *(Phoebis sennae)* 2¼ – 2¾"

The high-flying conspicuous Cloudless Sulphur is one of the largest of the sulphur butterflies in the United States. Its strong, straight, and rapid flight rarely includes a pause. Even though dispersals may reach Canada in some years, it is mostly widespread in the warm southern states. The southern California subspecies is named *marcellina*. The bright lemon-yellow male, slightly larger than the female, has minimal smudges on his wings; the female has many splotches and often adds subtle hues of pink, green, and orange to her wing coloration. Both sexes have a touching pair of silver spots on the underside of each hindwing. Adults sip nectar and mud while males patrol all day seeking females. The favorite caterpillar food plant is senna. Many caterpillars are green from eating senna leaves, but others are yellow from eating yellow senna petals. The unique chrysalis can be almost any color and has a distinctive angular shape held upright by a strand of silk.

10) Hermes Copper *(Lycaena hermes)* 1 – 1⅛"

Development and wildfires threaten the shrubland habitat of the Hermes Copper. This butterfly is not officially listed, but is definitely endangered for a number of reasons. The limited range includes western San Diego County with a few strays found in southern Riverside County and northern Baja California. The narrow flight period for this little tailed copper is around the month of June when the males perch on chaparral shrubs to await females. The sexes appear similar with the underwings yellow and big brown spots on the forewings. The upper hindwings are brown. On the spotted upperwings, the male has a wider brown border on the apex of the forewing. The adults nectar exclusively on the flowers of flat-top buckwheat and the caterpillars dine on only one buckthorn species, spiny redberry. The eggs hibernate until the next summer.

11) Great Purple Hairstreak *(Atlides halesus)* 1¼ – 1½"

The color purple is hard to find on the Great Purple Hairstreak. The Great Blue Hairstreak, an earlier name no longer used, seems more appropriate. The reason for the name change remains unknown. This large and stunningly beautiful insect has a black and orange abdomen with two long shiny blue tails on each hindwing. The underwings are a dull black with some iridescent blue; upper wings are iridescent blue with some dull black. Large spots of red are at the base of the wings. The adults are avid flower visitors and fly from early spring to late fall. The tiny green caterpillars eat mistletoe. In the afternoons, males locate little hills with small trees and shrubs to perch and wait for females.

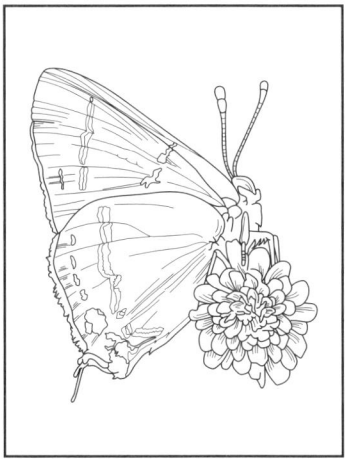

12) Gray Hairstreak *(Strymon melinus)* 1 – 1½"

The Gray Hairstreak is found all over the world. In fact an alternate name is the Common Hairstreak. The most prevalent hairstreak in North America is brownish gray above with fake eyespots and bogus antennae on the hindwings. The "eyes" are orange with a dark center; the "antennae" are really the hairstreaks. Below is pale gray with a dashed line of black, white, and orange from wing base to near the tip. The eyespots and the hairstreaks are prominent to allegedly divert would-be predators to the false head end. The abdomen is gray on females and orange on males. Unlike most hairstreaks the Gray basks with wings open and fly most of the year. The males perch on hills, trees, or shrubs seeking females from midday until dusk. The caterpillars will eat petals and leaves of many different plant species and this diversity of resources is one reason the Gray Hairstreak is so widespread.

13) Juniper Hairstreak *(Callophrys gryneus)* ⅞ – 1"

The Juniper Hairstreak belongs to a large complex of butterflies comprising eight or so subspecies including three in southern California. They resemble each other, especially on the underside, but live in different zones. In San Diego County the subspecies exist only fifty miles apart, but have different habitat requirements and caterpillar food choices. The widespread Nelson's Hairstreak meanders under coniferous forests in the mountains and the caterpillar munches on the scaled leaves of incense cedar. The rare Loki's Hairstreak roams the high desert and the caterpillar consumes juniper foliage. The very rare Thorne's Hairstreak flies only on Otay Mountain, near the Mexican border, where the caterpillar eats only the leaves of the equally rare Tecate cypress. Frequent wildfires are a threat to both the insect and the cypress. These Juniper Hairstreaks are very territorial and are easy to observe near their caterpillar food source. The males perch all day awaiting females.

14) Sonoran Blue *(Philotes sonorensis)* ¾ – ⅞"

The distinctive Sonoran Blue is almost neon in flight. The topside of this unmistakable butterfly is iridescent sky blue with bright red-orange patches. The butterfly's range includes the western side of the coastal mountains, eastern side of the Sierra, and desert edges from central California into central Baja California. The Sonoran prefers to flutter low on sunny slopes, rugged cliffs, and rocky canyons where Dudleya grows. The food plant has succulent leaves that are burrowed into by the variably-colored green to rose caterpillars. It has an early flight time beginning in January and continuing into early May. These butterflies were named for a gold rush area in California; there are no Sonoran Blues in mainland Mexico. Early butterfly naturalists wrote that this insect was "our most beautiful small butterfly."

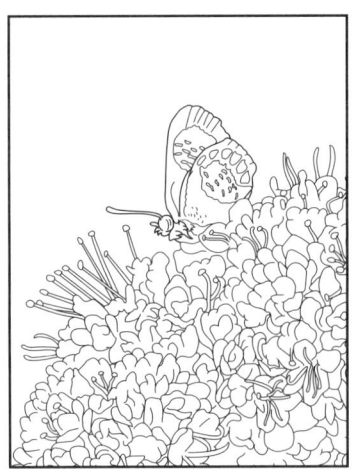

15) Western Pygmy Blue *(Brephidium exilis)* ⅜ – ¾"

The Western Pygmy Blue may be the smallest butterfly in the world at less than an inch across. Blue is not a dominant color on this tiny insect as it is mostly brown and gray with a small hue of blue at the base of the upperwings. On the underside of the hindwings are four large iridescent dots. It flies all year in southern California and other warmer locales. Habitat zones include lower altitude alkali flats, beach bluffs, and coastal salt marshes. It is a weak flier and stays close to the ground, but does journey on long northern dispersals to various environs and even into high mountain areas. The caterpillars eat the stems, leaves, flowers, and fruits of saltbush, tumbleweed, and other members of the goosefoot family.

16) Melissa Blue *(Lycaeides melissa)* ⅞ – 1¼"

One of the most prevalent blue butterflies in the west is the Melissa Blue. An older name, the Orange-Margined Blue, accentuates a special feature of an orange band on the trailing edge of all four wing-margins. Orange bands displayed by other blues will be solely on hind wing margins. It also includes striking iridescent spots on the underside. This handsome insect avoids the deserts and flies in coastal mountain meadows, marshy lake edges, and dry river bottoms. The green caterpillars chew on legumes, especially wild licorice, locoweed, and lupines while adults sip nectar and slurp mud. The males patrol all day seeking females. The eggs hibernate until the caterpillars emerge in the spring. Caterpillars of many blues, coppers, and hairstreaks have nectary glands that secrete a sugary substance that ants crave. Ants tend, protect, and defend the Melissa caterpillars that have these special glands.

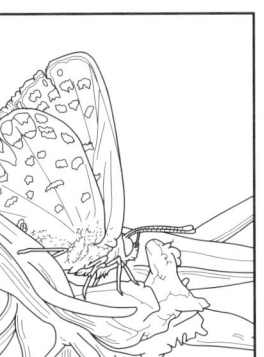

17) Acmon Blue *(Plebejus acmon)* ¾ – 1"

In the West, the Acmon Blue is common and widespread, especially in gardens and natural regions where buckwheat and wild legumes are available for the caterpillars to chew on their petals, pods, and leaves. The caterpillars are protected by ants that harvest the sugar in the nectary glands of the caterpillars. The male topside is powder blue and the female mostly brown with blue while the underside of both is pale gray. Near the trailing edge of the hindwings is a band of orange or pinkish ovals with noticeably shimmering iridescent blue dots. They avoid high alpine mountains but commonly visit foothills, open woodlands, forest openings, and even deserts. In southern California this insect normally flies from spring into fall, but may be seen any month of the year. The adults sip nectar and lap mud. The males patrol all day seeking females.

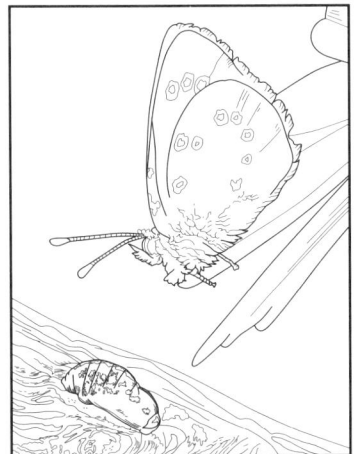

18) Silvery Blue *(Glaucopsyche lygdamus)* 1 – 1¼"

The Silvery Blue generally flies in open areas including shrublands, foothills, and mountain meadows from early spring through summer. The upperside is without markings but has a black edge with a white fringe. Males are blue; females are often brown. Both have a silvery-gray underside with a row of round black spots enclosed by white. Males patrol all day near legumes. Favorite foods for the caterpillars are lupines, deerweed, and other members of the legume family. The caterpillars can be almost any color and are attended by ants that harvest the sweet goo from their nectary glands. The Silvery Blue is very diverse and includes many subspecies. One subspecies, the Southern Blue (*australis*) is quite common in southern California. Another, the Palos Verdes Blue (*palosverdesensis*) was considered extinct until a small surviving population was discovered in San Pedro in 1994. A subspecies, now extinct, has an invertebrate conservation organization named after it: the Xerces Society.

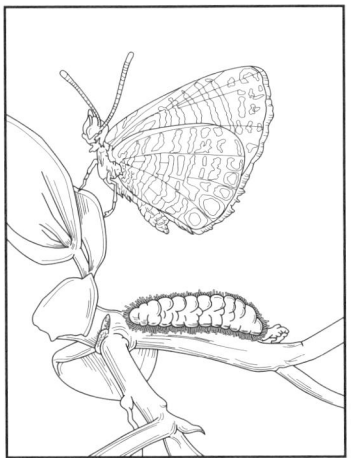

19) Marine Blue *(Leptotes marina)* ⅝ – 1"

The Marine Blue has a flight pattern more erratic than most other blue butterflies, and in southern California it flies almost all year. They are often found in urban neighborhoods near frequently planted shrubs of plumbago where the fuzzy-green caterpillars eat their blue flowers. In natural areas of open scrub and grasslands, and in agricultural fields, they eat the flowers of various wild legumes. Adults sip flower nectar and mud. On the underside of each squiggly-striped hindwing is a pair of glittering iridescent spots. An old out-of-use name, the Striped Blue, refers to this look of tan wavy lines. Males patrol all day on the lookout for females. Marine Blues may disperse as far north as Oregon and Minnesota and as far south as central Mexico.

20) Arrowhead Blue *(Glaucopsyche piasus)* 1 – 1¼"

The Arrowhead Blue is smoky blue with no markings on the top, but it has a black and white checkered fringe along the trailing wing edges. The gray underside has rows of black dots and patches of white "arrowheads" pointing inward. In southern California these blues fly in the spring while the males patrol all day seeking females near fields of lupine. This insect will fly in any meadow, foothill, or sage scrub area with substantial stands of lupine where the female lays her eggs. The caterpillars eat the flowers and fruits of lupines, locoweed, and a few other legume species. Ants tend the caterpillars in order to harvest the sugary liquid from their nectary glands. The Arrowhead's range in the West is barely into Baja California, but reaches as far north as Canada.

21) Mormon Metalmark (*Apodemia mormo*) ¼ – 1¼"

The Mormon Metalmark is a common and widespread species covering most of the western United States. It occupies many open environments including deserts, foothills, and woodland edges, and may be seen nearly every month of the year in warmer regions. The upperside is mostly orange with white spots and some black; the underside is mostly brown with white spots and some black. Metalmark flight is fast and erratic, but it still finds time to visit flowers, opening and closing its wings. Buckwheat florets are favorites for sipping nectar and the leaves are essential for laying eggs. Males perch on hillsides midday seeking females. The Mormon Metalmark complex is a group of butterflies with many subspecies. The southern California subspecies is called Behr's Metalmark (*virgulti*).

22) Coronis Fritillary (*Speyeria coronis*) 2 – 2¾"

The Coronis Fritillary is bright orange with black and white markings. The underside of the hindwings shows large black-rimmed, elongated ovals of iridescent silver, and therefore it is sometimes called the Coronis Silverspot. Habitats include oak and conifer woodlands, foothill chaparral, and mountain meadows. This insect is found during the warmer months in most of non-arid California and in the Great Basin. Eggs are not laid on plants. Near the end of summer, the eggs are haphazardly dropped near dried-up violets. The newly hatched caterpillars do not feed during winter, but wait until the plants leaf out in the spring when they feed at night. The adults sip nectar and mud. Males patrol all day long seeking females. The Coronis Fritillary subspecies in most of southern California is *semiramis*.

23) Gulf Fritillary (*Agraulis vanillae*) 2½ – 2⅞"

The topside of the Gulf Fritillary is a dazzling orange-red with black marks and tiny black-rimmed white spots on pointy forewings. It is sometimes called the Silver Spot because of the underside display of a dozen large silvery ovals that are striking in flight. It is found most months in the southern states and appears to be dispersing slowly northward. It occupies woodland openings, scrubby fields, and even deserts. This insect is also attracted to urban gardens, where passion vine, the caterpillar food plant, grows. Adults sip nectar preferably from red or white flowers; the males also lap mud. Flight is rapid and focused as the males patrol to seek females. Although the Gulf Fritillary will roost alone, it typically roosts in small groups of a dozen or so on a tree's lower leaves.

24) Variable Checkerspot (*Euphydryas chalcedona*) 1⅛ – 2"

The well-named Variable Checkerspot is part of a group with enormous geographical variation and nearly forty subspecies. It also includes closely related forms with many modifications in color and patterns. The butterfly's range is widespread, and it populates many environments in the western United States and even maintains a population in Alaska. Habitats include woodlands, riparian corridors, and brushy hillsides. The most common in southern California is the subspecies *chalcedona*, which has "checkered spots" on the topside that are mostly black with pale yellow and a little red. Other subspecies have more red or orange. The flight time is spring into summer. Favorite plants of the caterpillar include Indian paintbrush and monkeyflower, but other related plants are fed upon as well. Under harsh conditions the pupa can hibernate for several years. Males patrol hilltops or perch in canyon bottoms to encounter females.

25) Quino Checkerspot *(Euphydryas editha quino)* 1⅛ – 1⅞"

The striking Edith's Checkerspot includes three subspecies in southern California. The subspecies *quino*, found mostly in San Diego County, is highlighted since it is rare and endangered. It resembles many other checkerspots but has rounder forewings and no white on its black abdomen. In early spring, the Quino Checkerspot flies on open hills, canyons, and chaparral mesas seeking mates and the caterpillar food plant. The spiny black caterpillar consumes California plantain and a few other related plants. In explaining the decline in the insect's populations, one scientist noted that the butterfly was "routed out by development." The Quino inhabits areas sought by developers for condominiums, housing tracts, and shopping malls.

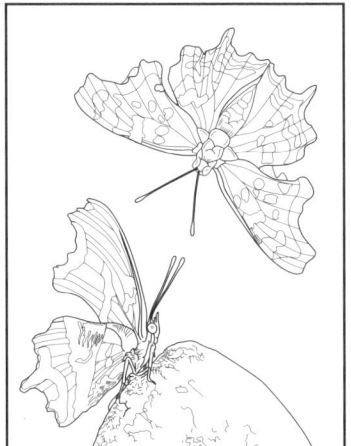

26) Satyr Comma Anglewing *(Polygonia satyrus)* 1¾ – 2"

The Satyr Comma is sometimes called Anglewing because it has uneven wing edges that are jagged and angled. The topside is golden-orange with dark dots, while the underside is brownish with a remarkable resemblance to tree bark or dead leaves. With wings closed, their camouflage allows them to nearly disappear on a tree trunk. Also, on the underside is a distinctive silver comma mark. The name "satyr" may reflect the woodland locale of the mythological man-beast who lived in "dark" places. Unlike its sun-loving cousins, the habitat of the Satyr Comma Butterfly is dark, shady riparian corridors. Nettle, the caterpillar food plant, also grows in that dank environment where the caterpillar sews leaves downward to make a little hideaway. This insect flies spring to fall in the western states seeking tree sap, rotting fruit, dung, and carrion rather than flowers. Adults hibernate until early spring but in southern California may fly on warm winter days.

27) Mourning Cloak *(Nymphalis antiopa)* 2⅞ – 3⅜"

The Morning Cloak is common all over the United States, Canada, Mexico, and many other parts of the world. It has interesting coloration including a maroon-brown topside with blue iridescent spots along a pale-yellow margin. The underside is similar. Environments reflect much diversity from arid lowlands to high mountains but must include meadow or streamside plants, particularly willow and cottonwood. The adults often hang out together and overwinter in a cluster. On a warm day in winter this butterfly may be seen flying around taking a pause from its mid-season slumber. They often add summer hibernation (estivation) to their strategy and may live up to a year. The spiky black caterpillars are gregarious and, as a massive alliance, may consume all the leaves of an entire elm tree within a week or so. Males sip mud, sap, dung, and carrion and perch in the afternoon to await females.

28) American Lady *(Vanessa virginiensis)* 1¾ – 2⅛"

The American Lady occurs all over the United States. Topside coloration is orange and black with white marks and a small identifying white dot on each of the front wings. The underside divulges an intricate cobwebby pattern and two large striking eyespots on each of the hindwings. Common habitats include open areas from coastal lowlands to the mountains where cudweed and pearly everlasting are eaten by the caterpillars. The caterpillar makes a nest of flower parts, leaves, and silk. All belonging to the genus *Vanessa*, the three lookalike "Lady" butterflies in southern California are the Painted Lady (most well known), West Coast Lady (most common), and the American Lady (least common).

29) Painted Lady *(Vanessa cardui)* 2 – 2¼"

The Painted Lady is a familiar butterfly on every continent and is perhaps the most widespread butterfly on the planet. Topside coloration is orange and black with white marks including four hindwing black dots often displaying blue centers. The underwing has four prominent eyespots in a cobwebby matrix. The adults are found in a variety of habitats where the nectar favorite is thistle and other asters. The caterpillar also prefers thistle, but will feed on most any plant. This ability to consume multiple food alternatives allows it to be successful all over the world. Fliers may be observed nearly every month of the year in southern California. In seasons of increased precipitation in Mexican deserts, huge swarms of Painted Ladies disperse northwestward through southern California and as far as Oregon. Between Scandinavia and Africa there is a Painted Lady migration similar to the famous Monarch epic between eastern United States and central Mexico. Obviously, the Painted Lady is a formidable and resilient flier.

30) West Coast Lady *(Vanessa annabella)* 1¼ – 2"

The West Coast Lady is the most common "Lady" butterfly in southern California, except in the years when throngs of Painted Lady butterflies disperse from Mexican deserts following an unusually wet year. The wing tips are squared off rather than pointed. Topside coloration is orange and black with white marks and a row of four tiny blue centered dots on the hindwing. The underside shows a web-like pattern and five assorted eyespots often obscured by the background. Habitats exclude most deserts but show great variation in wild areas as well as urban and suburban locales. Adults fly February to November but can be observed flying on warm winter days. The caterpillars consume mallows and nettles. As the name implies, the range of this butterfly in the United States includes most of the West.

31) Common Buckeye *(Junonia coenia)* 2 – 2½"

The Common Buckeye is easy to identify as it often rests on bare ground with brown wings open to display two pairs of orange bars and six huge eyespots on the upper wings. This insect is sometimes called the Peacock Butterfly because it shows prominent patterns that resemble the false eyes on the tail feathers of a strutting male peacock. Another name is the "King of the Road" as it occupies exposed fields, open scrubland, and regularly patrols trail sides and dirt roads. The caterpillar food plant includes plantain, monkeyflower, and related species. Since flower nectar is plain sugar water, the adult, like many other butterflies, supplements its sugary diet with salts, minerals, and other essentials found in mud, sap, dung, and carrion. Territorial males perch all day in dry gulches on low shrubs awaiting females and will confront other butterflies, challenge falling leaves, and sometimes surprise a tiny bird.

32) Lorquin's Admiral *(Limenitis lorquini)* 2¼ – 2¾"

Lorquin's Admiral is a delightful butterfly of southern California riparian corridors and moist forest openings from the coast to the foothills. This showy insect has a velvety dark brown color with a bold white, horizontal band across the topside, plus golden-orange patches on the wing tips. The underside is mostly brown and white. Also visiting similar habitats is the lookalike California Sister with orange wingtip patches surrounded by a black margin. The Lorquin's bizarre caterpillar is decorated with unique tubercles, bumps, spines, filaments, and horns but eats ordinary native willow and related plants. Caterpillars may overwinter in leaves rolled up with silk. Adults fly from spring to early autumn. Territorial males perch or patrol all day to encounter females. The assertive males may assail falling leaves, other butterflies, or occasionally a small bird.

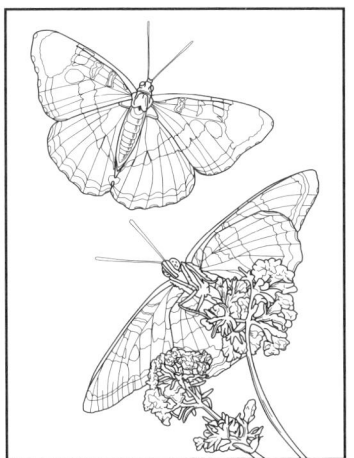

33) California Sister *(Adelpha bredowii)* 2⅞ – 3⅛"

The California Sister has an interesting name. It is not known if the moniker refers to a nun or a sibling. The black topside displays a broken white wing band, red bars on the leading edge, and an orange patch near each wingtip. The complex underside is similar but adds tints of gray, coffee, and lilac. The Sister is a high flier in the oak woodlands of California and non-desert regions of the Southwest but also visits riparian areas where the lookalike Lorquin's Admiral may be a mimic. The Sister seldom departs from the upper canopies but does journey to lower areas to mate and sample rotting fruit, mud, aphid honeydew, and occasionally nectar. The butterflies cruise spring to fall while the males patrol all day in gulches seeking females who will lay eggs on oaks. The resilient caterpillars hibernate until spring.

34) Monarch *(Danaus plexippus)* 3⅛ – 4"

The attractive Monarch is probably the most familiar butterfly in North America and possibly the most famous in the world. The bright-orange male is easily recognized with thin black veins and white dots on the wing margins. He also has a small scent patch on each inner vein. The female has thicker veins and lacks the scent patches. They occupy almost any environment that has milkweed nearby and are seen all over the United States, southern Canada, and Mexico. In autumn the Monarchs east of the continental divide journey into the high mountains of central Mexico to overwinter. In the spring they return, repopulate throughout the summer, and repeat the fall dispersal after several generations. The butterfly population west of the divide has a similar epic, but they overwinter along the California coast until they return inland to repopulate the western states. During the winter months, the adults drink only water and rest. Non-migrating butterflies breed year-round in the warmer regions. Monarch caterpillars eat only milkweed.

35) Monarch Caterpillar *(Larva)* and Chrysalis *(Pupa)*

Within a week after the egg is laid on a milkweed plant the little caterpillar emerges, consumes the protein-rich egg case, and then begins to munch on nearby leaves. As it matures the tiny black caterpillar sheds its skin a number of times and progresses into a more colorful larva by adding bands of black, white, and yellow and growing a pair of filaments on each end. Inside of two weeks, the caterpillar locates a safe site, attaches itself to a twig or leaf and changes into the pupa (chrysalis). Alterations occur inside the chrysalis to transform a plant eating, worm-like crawler into a nectar-sipping, winged creation quite different than what it was before. In about two weeks, the adult will emerge from the chrysalis with wet wings to dry and mouthparts to align into a feeding tube (proboscis). Within hours it is ready to fly and look for nectar, milkweed, and a mate. The adult monarch may live another six weeks unless it is the generation that migrates and then it may live six months or more.

36) Western Monarch Clusters

The classic migration of the eastern Monarch begins around the autumnal equinox in the eastern United States and Canada as the days grow shorter and nights turn cooler. About five million Monarch Butterflies prepare for an epic journey and head south funneling through Texas and traveling up to 100 miles per day to end up in central Mexico on a dozen fir-forested mountaintops to spend the winter in giant clusters. In spring the Monarchs fly north where their descendants repopulate the Midwest until the autumn generation repeats the legendary migration with new squadrons of orange fliers. Monarch totals west of the continental divide are much less than a quarter million and their numbers are dwindling. They come from the western mountains in the fall to overwinter in clusters within the eucalyptus groves and coniferous forests along the California coast from just south of San Diego to just north of San Francisco. In the spring they mate and return to mountain meadows where future generations will repopulate the West until the equinox signals the cycle to begin again.

37) Queen *(Danaus gilippus)* and Monarch *(Danaus plexippus)* 3 – 3⅜" | 3⅛ – 4"

The Queen (bottom) is frequently misidentified as the closely related Monarch (top) because of similar coloration and comparable wing patterns especially on the underside. The Queen does not migrate like the legendary Monarch, but does roost in much smaller clusters like it's cousin and disperses into northern regions in the summer. The insect flies spring to autumn in the Southwest and southeastern coastal regions and all year where temperatures are moderate. The butterfly may occupy most any open habitat including gardens and especially deserts. Caterpillars of the Queen feed only on milkweed and, like the Monarch, sequester milkweed toxins that transfer into the adults to make them distasteful to many predators. The caterpillars look similar except the Queen possesses three pairs of filaments instead of two pairs like the Monarch. The jade chrysalids are indistinguishable. The Queen has chestnut-brown on the topside with white dots rather than the Monarch's bright orange with dark veins. Underside hindwing venation is nearly identical on both of these delightful butterflies.

38) Skippers: [Fiery *(Hylephila phyleus)*, Juba *(Hesperia Juba)*] 1¼ – 1½"

One third of all North American butterflies belong to a group called the skippers. They have stout bodies, wide heads, and small wings but are still fast fliers. Skippers also have short antennae with hooked clubbed ends, large eyes, and a long proboscis to efficiently sip flower nectar, water, and mud. To some they resemble moths. Two familiar southern California Skippers are the Fiery Skipper and the Juba Skipper. Prior to take off they both display a "jet-plane" posture with hindwings more open than the forewings, characteristic of moths. Orange-brown and black are their dominant colors with the underside of the Fiery Skipper showing "measles-spots" and the Juba presenting white angular marks. They fly from spring to fall in natural regions of chaparral and grasslands, and in urban neighborhoods they visit lawns and weedy roadsides where the caterpillars of both species feed on various grasses.

39) Ceanothus Silk Moth *(Hyalophora euryalus)* 3½ – 5"

The Ceanothus Silk Moth belongs to the wild silk moth family (Saturniidae) and does not produce textile silk. Domesticated moths of another family (Bombycidae) produce commercial silk. This reddish brown moth has white markings including a "Nike-swoosh" and wingtip eyespots. After the Silk Moth emerges from a golf-ball sized cocoon, it flies along shrubby hillsides from late winter to midsummer with non-functioning mouthparts. These adult insects cannot feed, but will live long enough to reproduce and for the female to glue eggs to chaparral twigs. The bizarre green caterpillar with black spikes on red and blue tubercles consumes the leaves of Ceanothus and many other woody shrubs. Moths are not butterflies, but belong to the same insect order, Lepidoptera. Moths have thick bodies, drab wings, usually fly at night, have diverse antennae, and their caterpillars often construct a cocoon around the chrysalis. Butterflies have thin bodies, colorful wings, are day fliers, have clubbed antennae, and their caterpillars do not build a cocoon. Over 90% of Lepidoptera species are moths.

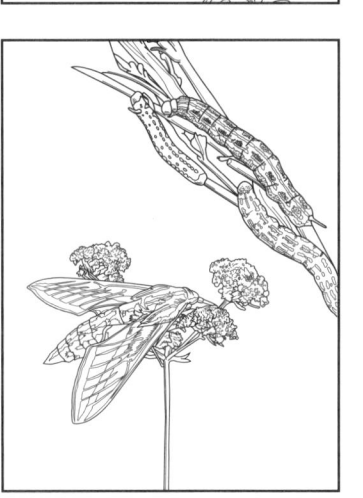

40) White-lined Sphinx Moth *(Hyles lineata)* 2½ – 3½"

The White-lined Sphinx Moth is dark brown with a white line between wingtips. From early spring to late fall at dawn or dusk it hovers from flower to flower, sipping nectar, just like a hummingbird. This insect flies in many habitats from the coast to the mountains and into the deserts. Caterpillars eat mostly members of the evening-primrose family. Unlike many other moth species, the caterpillar will burrow into the soil and produce a chrysalis with no cocoon. In years of ample rainfall a super-bloom of desert plants in Anza-Borrego and other southern California deserts occurs. What often follows is huge populations of newly emerged caterpillars devastating the flower fields. Migrating Swainson's Hawks reduce the multitudes as they fly through the arid regions feeding on the crawling critters. The variously tinted caterpillars are called hornworms because of an imposing, but harmless barb protruding from atop their posterior. The name sphinx is derived from the behavior of other sphinx moth caterpillars that when alarmed arise and resemble an Egyptian statue.

Introducing the **Color & Learn** series of informational coloring books from Sunbelt Publications. Every book features interpretive information for each engaging line drawing. Choose from a variety of unique subjects.

•

Coloring Southern California Butterflies & Caterpillars
9781941384596 | $10.95
Bill Howell

Coloring Lizards, Snakes & More: Southern California
9781941384558 | $9.95
Bradford D. Hollingsworth

Coloring Southern California Birds
9781941384473 | $10.95
Wendy Esterly

Coloring Nature in the California Chaparral
9781941384459 | $9.95
Richard W. Halsey

Coloring California Missions
9781941384381 | $9.95
Max Kurillo

Coloring Plants Used By Desert Indians
9781941384374 | $9.95
Diana Lindsay

***Coloring Who-o-o's Awake in the Desert
with Word Searches***
9781941384350 | $9.95
Jenny Holt
Companion to *Who-o-o's Awake in the Desert*
9781941384312 | $12.95

***Coloring Metal Sculptures
The Magical Works of Ricardo Breceda***
9781941384343 | $9.95
Diana Lindsay

Coloring San Diego Landmarks
9781941384251 | $9.95
Nancy Hendrickson

San Diego, CA
(619) 258-4911 • (800) 626-6579
www.sunbeltpublications.com
info@sunbeltpub.com

Coloring Books That Educate As You Create

More than just a coloring book, titles in the Sunbelt Publications *Color & Learn* series contain interpretive text and a variety of informative graphics including charts, photographs, maps, and diagrams, making them perfect for learning at home.

Interpretive Text • Photographs • Charts • Maps • Diagrams

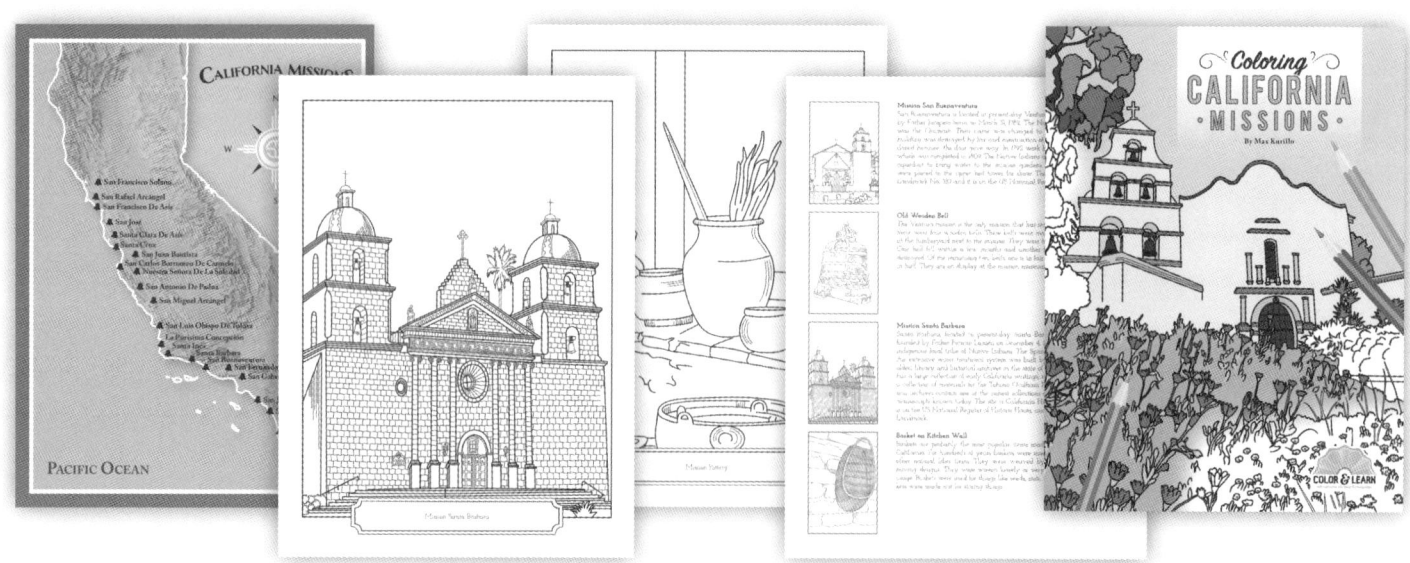

Have Fun Learning!

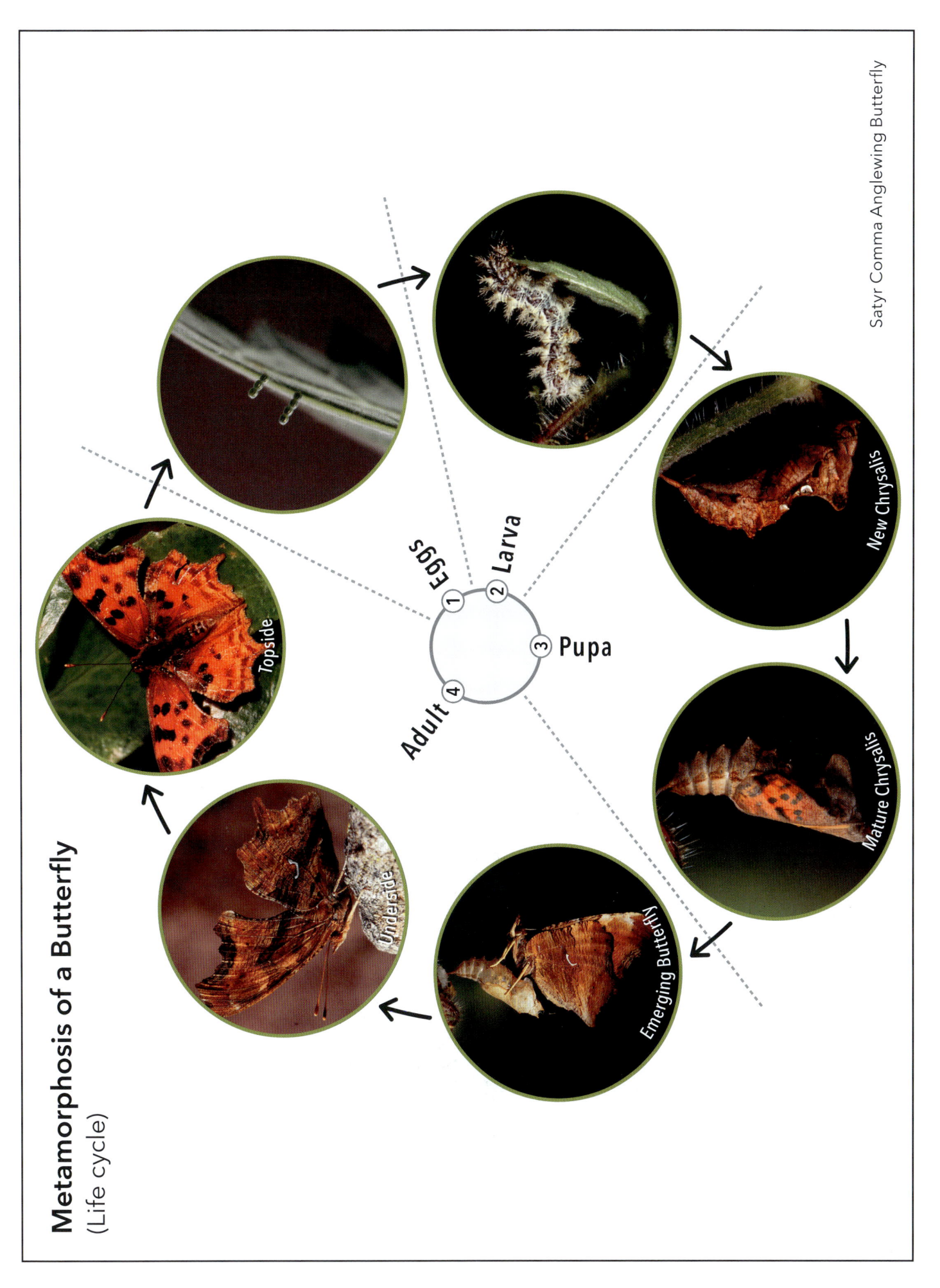

PARTS OF A BUTTERFLY

HEAD = H (1 Segment)

H1. ANTENNA (2)
All butterflies have a pair of clubbed antennae for detecting scents, temperatures, and breezes.

H2. COMPOUND EYE (2)
Each compound eye has thousands of little lenses especially sensitive to movement, but also can detect color and blurry images.

H3. COILED PROBOSCIS (1)
Butterflies have a long straw-like mouthpart that is used to suck up liquids like water, nectar, and sap.

THORAX = T (3 Segments)

T1. FOREWING (2)
The topside often has colors and patterns that identify it from the opposite sex.

T2. HINDWING (2)
The underside often has colors and patterns to camouflage it from enemies.

T3. LEG (6)
Butterflies have six legs. Some have a very tiny front pair and appear to have only four legs. They can taste with their feet so a female can lay eggs on the correct plant for the caterpillar.

ABDOMEN = A (10 Segments)

A1. ABDOMEN (1)
It has ten flexible segments containing organs for digestion, respiration, and reproduction.

A2. SPIRACLES (18)
Holes for air exchange. Two pairs on thorax, 7 pairs on abdomen.

A3. REPRODUCTIVE STRUCTURES (sexes separate)
At the end of the abdomen is an egg-laying tube in females and a sperm transferring organ in males.

PARTS OF A CATERPILLAR

HEAD = H (1 Segment)

H1. SIMPLE EYE (12)
Sees UV light and some color.

H2. ANTENNA (2)
Minimal function in caterpillar.

H3. MANDIBLE (2)
For chewing plants.

H4. SILK GLAND (1)
Produces silk for building leaf nests, spinning pads for attaching chrysalids, and for moths to construct cocoons.

THORAX = T (3 Segments)

T1. THORACIC LEG (6)
Minimal use in caterpillar.

ABDOMEN = A (10 Segments)

A1. PROLEG (10)
To cling with. Last pair may become structure to attach chrysalis.

A2. SPIRACLE (18)
Holes for air exchange. One pair on thorax, 8 pairs on abdomen.

Anise swallowtail

Anise Swallowtail
(*Papilio zelicaon*)

Anise Swallowtail
(*Papilio zelicaon*) Larva

Anise Swallowtail
(*Papilio zelicaon*) Larva

Giant Swallowtail
(*Papilio cresphontes*)

Giant Swallowtail
(*Papilio cresphontes*) Larva

Pale Swallowtail
(*Papilio eurymedon*)

Pale Swallowtail
(*Papilio eurymedon*) Pupa

Western Tiger Swallowtail
(*Papilio rutulus*)

Western Tiger Swallowtail
(*Papilio rutulus*) Pupa

Checkered White
(*Pontia protodice*)

Sara Orangetip
(*Anthocharis sara*)

Orange Sulphur
(*Colias eurytheme*)

California Dogface
(*Colias eurydice*)

Cloudless Sulphur
(*Phoebis sennae*)

Cloudless Sulphur
(*Phoebis sennae*) Larva

Cloudless Sulphur
(*Phoebis sennae*) Pupa

Hermes Copper
(*Lycaena hermes*)

Great Purple Hairstreak
(*Atlides halesus*)

Gray Hairstreak
(*Strymon melinus*)

Juniper Hairstreak
(*Callophrys gryneus*)

Sonoran Blue
(*Philotes sonorensis*)

Western Pygmy Blue
(*Brephidium exilis*)

Melissa Blue
(*Lycaeides melissa*)

Melissa Blue
(*Lycaeides melissa*)

Acmon Blue
(*Plebejus acmon*)

Silvery Blue
(*Glaucopsyche lygdamus*)

Silvery Blue
(*Glaucopsyche lygdamus*) Pupa

Marine Blue
(*Leptotes marina*)

Marine Blue
(*Leptotes marina*) Larva

Arrowhead Blue
(*Glaucopsyche piasus*)

Arrowhead Blue
(*Glaucopsyche piasus*) Eggs

Mormon Metalmark
(*Apodemia mormo*)

Coronis Fritillary
(*Speyeria coronis*)

Gulf Fritillary
(*Agraulis vanillae*)

Gulf Fritillary
(*Agraulis vanillae*)

Variable Checkerspot
(*Euphydryas chalcedona*)

Quino Checkerspot
(*Euphydryas editha quino*)

Quino Checkerspot
(*Euphydryas editha quino*) Larva

Satyr Comma Anglewing
(*Polygonia satyrus*)

Satyr Comma Anglewing
(*Polygonia satyrus*)

Mourning Cloak
(*Nymphalis antiopa*)

American Lady
(*Vanessa virginiensis*)

American Lady
(*Vanessa virginiensis*)

Painted Lady
(*Vanessa cardui*)

Painted Lady
(*Vanessa cardui*)

West Coast Lady
(*Vanessa annabella*)

West Coast Lady
(*Vanessa annabella*)

Common Buckeye
(*Junonia coenia*)

Lorquin's Admiral
(*Limenitis lorquini*)

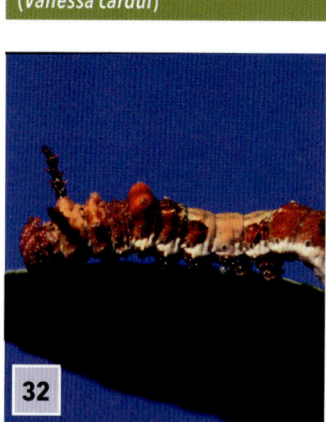
Lorquin's Admiral
(*Limenitis lorquini*) Larva

California Sister
(*Adelpha bredowii*)

California Sister
(*Adelpha bredowii*)

Monarch
(*Danaus plexippus*) F

Monarch
(*Danaus plexippus*) M

Monarch Caterpillar
Larva

Monarch Chrysalis
Pupa

Monarch
(*Danaus plexippus*)

Western Monarch Clusters

Skipper
Fiery (*Hylephila phyleus*)

Queen
(*Danaus gilippus*)

Skipper
Juba (*Hesperia Juba*)

Ceanothus Silk Moth
(*Hyalophora euryalus*) Larva

Ceanothus Silk Moth
(*Hyalophora euryalus*)

White-lined Sphinx Moth
(*Hyles lineata*) Larva

White-lined Sphinx Moth
(*Hyles lineata*)

Photo by: Rick Halsey

Bill Howell's love of butterflies and other small invertebrates has been passed on to many from his decades of teaching students of biology and related sciences in local universities, community colleges, and high schools. In 1985 he became the lead instructor for the Canyoneers of the San Diego Natural History Museum and since 1995 also head trainer for the Trail Guides of Mission Trails Regional Park.

Bill was past president of the Monarch Program in Encinitas, California, a research and educational organization involved with tagging and counting overwintering monarchs along the California coast. He has also led eco-tourism visits to the migrating monarch's magical clustering sites in the mountains of central Mexico. Bill is often asked to give talks and share his knowledge and passion for these frequent fliers and related topics to interested groups.

As an award-winning photographer, Bill's images have captured the essence of these enchanting creatures and hopes to encourage members of new generations to become certified butterfly fanatics.